Why the Lion Grew Its Mane

LEWIS SMITH

Why the Lion Grew Its Mane

A MISCELLANY OF RECENT SCIENTIFIC DISCOVERIES **FROM ASTRONOMY TO ZOOLOGY**

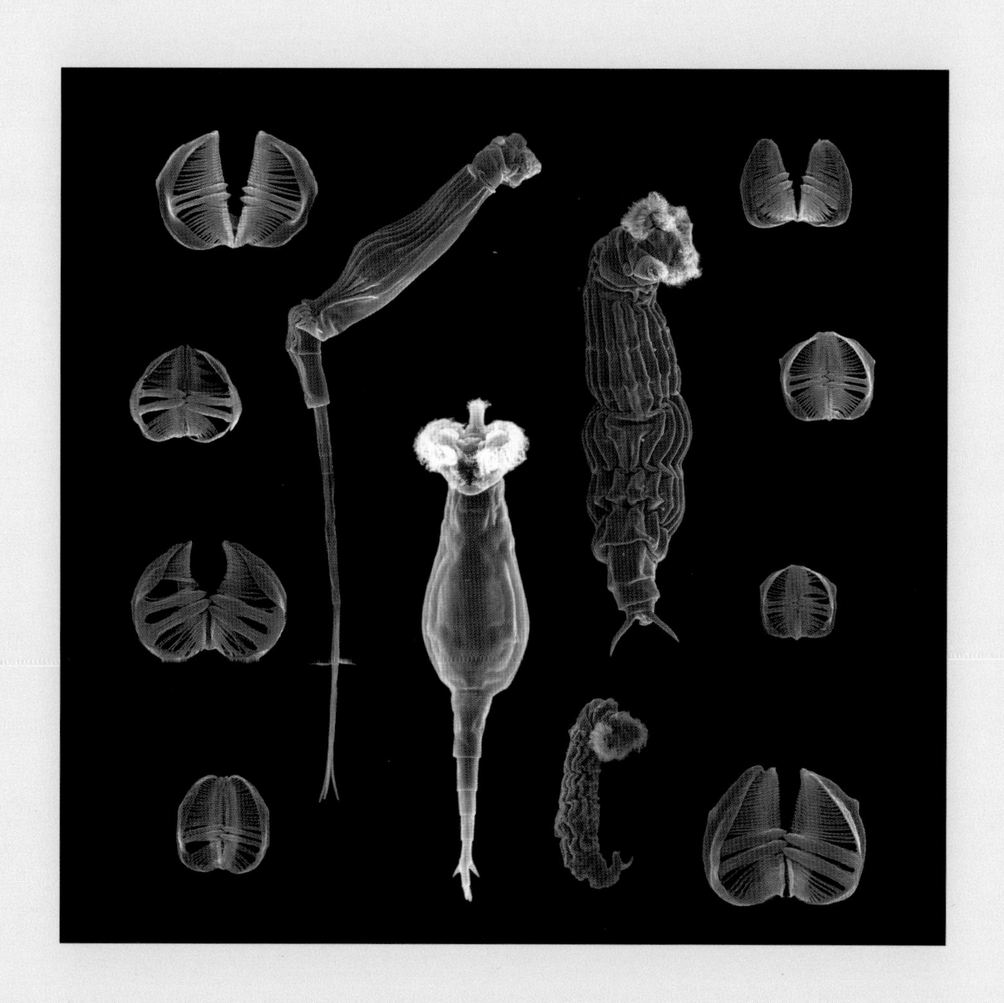

PAPADAKIS

Design Director: Alexandra Papadakis
Design assistant: Shirlynn Chui
House Editor: Diana Moutsopoulos
Research assistant: Hayley Williams

First published in Great Britain in 2007 by
PAPADAKIS PUBLISHER

PAPADAKIS

An imprint of New Architecture Group Ltd.

Kimber, Winterbourne
Berkshire, RG20 8AN
www.papadakis.net

ISBN: 978 1 901092 87 3

A CIP catalogue of this book is available from
the British Library.

IMAGE CREDITS:
half title: Chimpanzees gesturing [photos courtesy of Frans de Waal / Living Links Center] see page 250

title verso: An older African lion with a thick mane [photos: James Warwick / NHPA] see page 20

title page: Scanning electron micrographs showing variations of bdelloid rotifers and their jaws [image courtesy of Diego Fontaneto] see page 142

overleaf: Stromatolites photographed in Shark Bay, Western Australia [photo: Georgette Douwma / Science Photo Library] see page 122

page 8: view of CERN's CMS detector in the surface hall at Cessy [photo courtesy of Maximilien Brice; © CERN] see page 78

contents: The break up of the B15 iceberg, Antarctica [photo courtesy of Jacques Descloitres, MODIS Land Science Team. Image courtesy of NASA] see page 262

For Heather and Willow, with love

Acknowledgments

Like many other lifelong readers, I have long harboured a desire to write a book without ever really expecting to get such a project off the ground. That it has now happened is thanks in large part to the efforts of others and I should like to acknowledge some of them here. Chief amongst them are my publishers, Andreas and Alexandra Papadakis. The idea for the book was theirs and without their confidence and vision it would never have been written. Furthermore, it is thanks to Alexandra's eye for layout and design that the book looks as attractive as it does. I am also immensely grateful to Diana Moutsopoulos for her work on editing and obtaining pictures.

Scientists around the globe are working hard to unravel the world around us. Without them and their curiosity and dedication, answers to life's mysteries would be so much rarer and, like so much else, this book would have been impossible without their discoveries and insights. Many scientific discoveries are first reported in peer-reviewed journals or specialist publications. These have been a source of information and inspiration for this book. They include: *Science, Nature, Proceedings of the Royal Society A, Proceedings of the Royal Society B, Proceedings of the National Academy of Sciences, The Astrophysical Journal, Journal of Clinical Investigation, Evolution and Human Behaviour, Accident Analysis & Prevention, Current Biology, Astrobiology, Journal of Zoology, Genome Research, New Journal of Physics, Nature Genetics, Nature Materials, Molecular Ecology Notes, PLoS Biology, Pediatrics, Journal of Experimental Biology, Geophysical Research Letters, Psychological Science, Astrophysical Journal Letters, Biology Letters, European Journal of Human Genetics, Neurology*, and *Zootaxa*. A good number of scientists have taken the time to check my facts where they relate to their research. I am grateful to them for their observations and accuracy.

The Times newspaper has given me the wonderful opportunity to write about science and the environment, for which particular thanks must go to John Wellman, Mark Henderson, Ben Preston and Robert Thomson.

I would like to thank my parents, Pat and Harvey Smith, for their support and encouragement, and most especially my wife, Heather Chinn, for her endless patience, encouragement and help in writing this book.

Lewis Smith
September 2007

SHAWTON
ME&RE+1/3/04
ME&RE+1/3/02
ME&RE+1/3/01
ME+1/1/03
ME+1/1/02
ME+1/1/01
ME+1/1/36

Fp1 -F3
F3 -C3
C3 -P3
P3 -O1
T7 -T3
T3 -T5

Contents

previous, clockwise from top left: Chimpanzee [photo: John Foxx / Getty], trap-jaw ant [photo: Alex Wild], wild Komodo dragon [photo courtesy of Richard Gibson, Zoological Society of London], two *Panthera leo* males resting [photo: Jonathan & Angela Scott /NHPA], Magellanic penguins [photo courtesy of Rory Wilson, University of Wales, Swansea]

Animal Behaviour

Virgin births, animal shopping trips and tiny creatures with the fastest bites in the world are just some of the often bizarre secrets nature has revealed to science.

When naturalists in England were first sent the pelt of a platypus, they thought the creature so outlandish that they entertained strong suspicions it was a hoax. Unscrupulous taxidermists had already tried to fool the world into believing in mermaids by dint of stitching the top halves of monkeys to fish tails—so a creature with fur, webbed feet, a beaver-like tail and a duck-like bill was always going to raise eyebrows.

The platypus, also known as the duck-billed platypus, turned out to be entirely real, but the animal kingdom has continued to provide a succession of surprises. Many creatures that have long been officially classified are still slow in giving up their secrets. After all, it wasn't until the 1880s, almost a century after European scientists learned of the existence of the platypus, that they confirmed it laid eggs.

The immaculate dragon

It took almost a century after Komodo dragons were first documented in 1910 for it to be discovered that the females can have offspring through virgin births.

Despite the size of the world's largest lizards, which can grow up to 10 feet (3 m), no one had realised they could breed without mating until a female called Sungai produced eggs at the London Zoo some two years after she was last known to have been in contact with a male. Suggestions that Sungai had somehow managed to store the sperm of Kimaan, the last male she lived with, were dismissed when genetic tests proved that her four offspring were the result of parthenogenesis.

Parthenogenesis derives from the Greek words for virgin and birth, and occurs when an unfertilised egg begins dividing as if it was an embryo. All offspring from this type of asexual reproduction in Komodo dragons are male. In humans a female has two X chromosomes and a male has one X and one Y. Komodo dragons, however, have W and Z chromosomes and the female has the mix of two. When a female komodo lays an egg through parthenogenesis, she passes on just the one chromosome, which is then duplicated. While a WW combination is unviable, a ZZ combination results in a male.

The DNA tests carried out by Phillip Watts of the University of Liverpool in the UK showed that Sungai's offspring were not clones, but that their genetic makeup was derived solely from one female. Analysis similarly showed that three failed eggs from a clutch of 11, laid by a Komodo dragon at Chester Zoo in the UK, were also produced through parthenogenesis. The mother, Flora, had never had a mate.

The discovery that Komodo dragons can reproduce without mating shows it would be possible for a female dragon washed up on an island to keep her lineage going—first through parthenogenesis and later by breeding with her sons. It does, on the other hand, present a problem for conservationists anxious to maintain a gene pool that is as large and varied as possible while running breeding programmes. The finding also highlights how little rather than how much we understand about many species, as well as the intricacies and balance of the natural environment.

Sungai's offspring were not clones, but their genetic makeup was derived solely from one female

right: Wild Komodo dragons [photos courtesy of Richard Gibson, Zoological Society of London]
opposite: The first parthenogenetic dragon hatching at Chester Zoo [photo courtesy of Dr Phill Watts, University of Liverpool]
overleaf: Komodo dragon, Indonesia [Photo: NHPA / Jonathan & Angela Scott]

Baldness as

a sign of virility

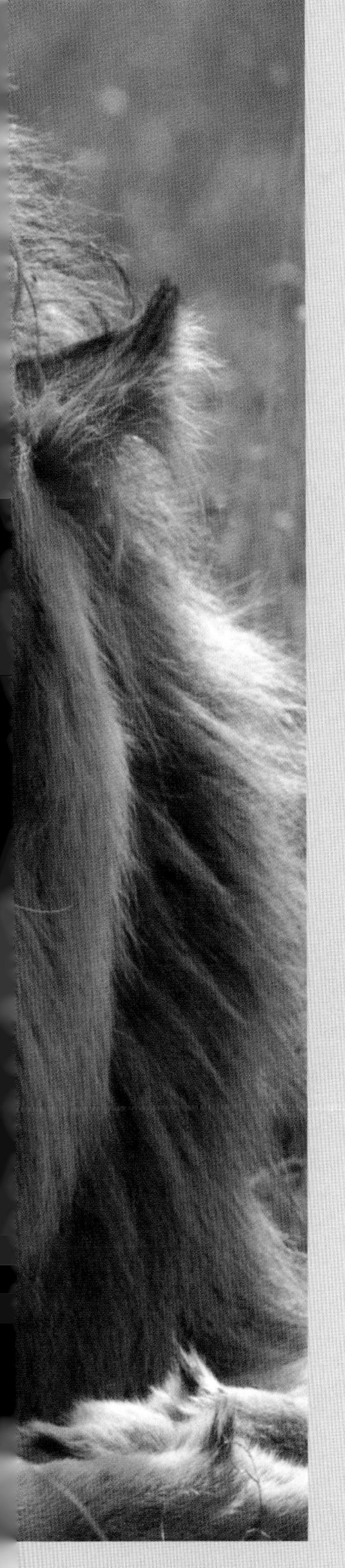

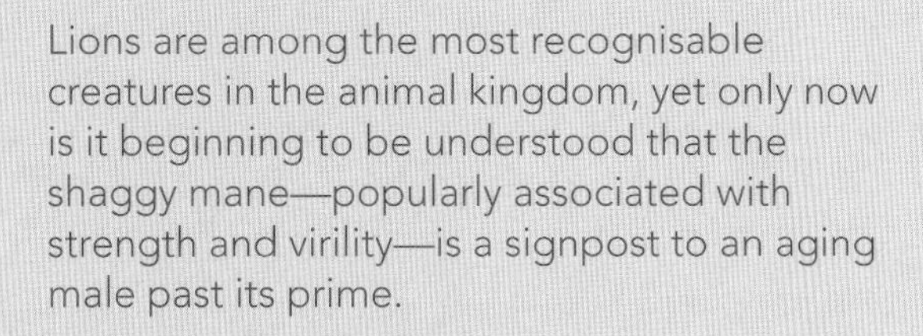

Lions are among the most recognisable creatures in the animal kingdom, yet only now is it beginning to be understood that the shaggy mane—popularly associated with strength and virility—is a signpost to an aging male past its prime.

Professor Julian Kerbis Peterhans of Roosevelt University and the Field Museum in Chicago, US, was part of a team that challenged the popular assumption that a male African lion is at its peak when it has an impressive mane. Whereas a male lion is generally in its prime for breeding from about five.to seven years old, the researchers found that the animals that sport the thickest and most obvious manes are the older specimens, those less likely to attract a female.

Professor Kerbis Peterhans said the seven-year study of wild lions showed that their manes continue growing well past their best years, so that in old age they appear, at least to us, to be at their most magnificent. To the lionesses, however, a thick mane indicates that the male is probably over-the-hill and lacks the animal magnetism required of the perfect mate. Far from advertising strength and virility, ample manes warn females that they might be better off looking elsewhere for a father for their cubs.

Moreover, because the longer-maned males are older, they are less likely to be the dominant male of the pride—a more youthful and vigorous lion having taken charge—and therefore are not as appealing to a female.

The research team's studies also found a correlation between mane thickness and climate, with the manes becoming thinner the hotter and more humid the weather. In each of these climate ranges the age rule applied, even to the supposedly maneless lions of the greater Tsavo ecosystem near the equator in Africa, which the research team found in 87 per cent of cases do in fact have a mane when the males reach maturity, albeit a comparatively thin and wispy one.

Professor Kerbis Peterhans said his team observed the pattern of shorter manes for fitter males among all African species. This included recently extinct types, for which researchers found evidence from museum specimens.

Far from advertising strength and virility, ample manes indicate that a male lion is past its prime

left: Two young *Panthera leo* males resting in Kruger National Park, South Africa
[photo: Jonathan & Angela Scott / NHPA]

left: A young male African lion with a scant mane, indicating he is in his reproductive prime
right: An older African lion with a thick mane. Lions with more substantial manes are typically past their reproductive prime [photos: James Warwick / NHPA]

Chimps and the allure of wrinkles

Male chimps find sagging skin, wrinkles, stretched nipples and bald patches on female chimps exciting badges of honour

An active preference for geriatric mates among male chimpanzees has been identified in a study of apes at the Kibale National Park in Uganda. Sagging skin, wrinkles, stretched nipples and bald patches in female chimps are anything but a turn-off for the males, instead regarding the signs of age as exciting badges of honour.

Male chimps want their offspring to be given the best start in life, and to that end they find nothing more attractive than an elderly female who has had years of experience in bringing up youngsters.

In the eyes of a male chimp, the well-toned body of a young female with all her fur is merely a sign of inexperience both in bringing up young chimps and surviving herself, according to research led by Dr Martin Muller of Boston University in the US. So important is an aging partner to the males that fights often break out between them over rights of access to the oldest females.

Dr Muller said the impulse in male chimps to be toyboy to an aging female is so strong that they disdain the younger female chimps. Perfectly-formed young females generally have to make do with the geekiest males of the group.

The observations of chimp behaviour and preferences allowed the researchers to suggest that the tendency in humans to equate female youth with beauty developed independently after man's lineage had evolved from a common chimp-like ancestor.

Human ideals of beauty are likely to have been driven in part by the existence of the menopause in women, as men would know that the choice of a young partner maximises the years available to have children. Female chimps do not go through the menopause and thus avoid such a limitation on the number of years they are fertile. Equally, chimps are not monogamous, which means the males, the study suggests, need only concern themselves with the likelihood of their chosen partner surviving long enough to have and raise one baby rather than several children.

below: Chimpanzee hand [photo: Image Source White / Getty]
opposite: Chimpanzee [photo: John Foxx / Getty]

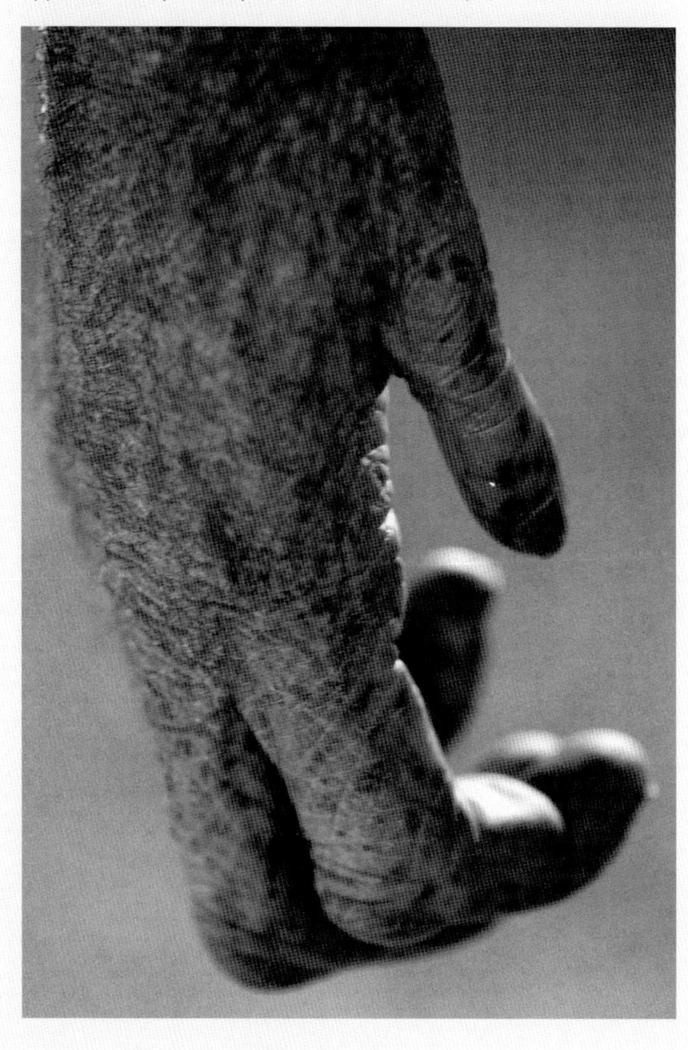

School for meerkats

Chimps and other apes that use tools are thought to learn how to use them by watching their mothers or other members of the family and then trying for themselves. The art of cracking nuts with stones is thought to take up to seven years to master.

The extent to which the young are taught as opposed to learning by observation remains unclear, but active schooling does take place for at least one animal—the meerkat.

Meerkats, a type of mongoose that live in the Kalahari Desert of southern Africa, became the first creature proven to tutor the young in their groups as part of a study by the UK's University of Cambridge. By going to the effort of teaching the young, through a technique called opportunity learning, the older meerkats are believed to speed up the pups' learning process.

Older members of the group, and not just the parents, were seen to seek out the pups to help teach them what was edible and what was dangerous. Like any good teacher, the helpers continued to monitor the way the pup coped when an item of prey had been passed over for it to eat.

If the pup showed reluctance to tackle the prey speedily, the older meerkat would nudge it with nose or paws towards its pupil as an act of encouragement. If the prey managed to scuttle or creep away from the pup, the older animal would fetch it back and sometimes disable it.

Just as children find at school, once the pups had mastered one task, the adults would challenge them to try something harder. For the first lesson the pups would be given the simple task of grasping and eating a dead insect or grub. As they became adept at that, the older meerkats would provide them with live but injured meals. Later lessons would involve fully mobile food or even more dangerous prey, such as scorpions with their stingers removed.

For the first lesson the pups were given the simple task of grasping and eating a dead insect or grub

above: Meerket sentinels in Kalahari, South Africa, keep watch at sunset [photo: Martin Harvey / NHPA]
opposite: Meerkets in Kalahari, South Africa [photo: Nigel J Dennis / NHPA]

Warbling for rappers

The great tit is a bird that has attracted the attention of researchers for its singing voice rather than its mental processes. It has been discovered that, in cities across Europe where great tits are found, they sing much more raucously than their rural cousins.

To be heard over the noises of cars, trains, machinery and the other city hubbub, the urban male tits have learned to perform a type of avian rap. Discarding the traditional and more melodious calls heard in the countryside, they have experimented with pumping out staccato notes at higher pitches.

Though their study was limited to 10 major cities, including London and Rotterdam, researchers from the University of Leiden in the Netherlands believe it is likely that the birds' calls have changed in all the noisy urban areas in which they live. They said that the calls are now so different that urban great tits may eventually have to be reclassified as a distinct subspecies.

The overall effect of the change in singing style is shorter and sharper calls, especially for the introductory notes, with a greater level of innovation. Traditional great tit tunes comprise of two, three and four notes. The urban birds invented many more single-note and five-note calls, all sung much faster than the rural tunes. When five-note tunes were sung, they were completed in less time than the more melodious calls. One bird in Rotterdam, probably imitating a blue tit, tried out a 16-note variation but did so at top speed.

Changes in call patterns seemingly apply to both territorial and mating calls. In cities, with the constant backdrop of human-produced noise, louder mating calls are needed, quite simply, to ensure a potential mate can hear the call and be attracted. Sharper territorial calls help avoid bloodshed, as they reduce the chance of a bird wandering unwittingly on to another's turf.

The discovery that urban great tits are innovating their calls is in many ways an optimistic sign for conservationists, because it demonstrates a degree of adaptability that could not have been easily anticipated.

One bird in Rotterdam tried out a 16-note variation at top speed

above: Great tit pictured in London
[photo courtesy of Philip Greenspun]
right: Sonogram used in the University of Leiden's research comparing call patterns of urban and rural great tits
[courtesy of Hans Slabbekoorn]
opposite: Great tit, *Parus major* [photo: VIREO]

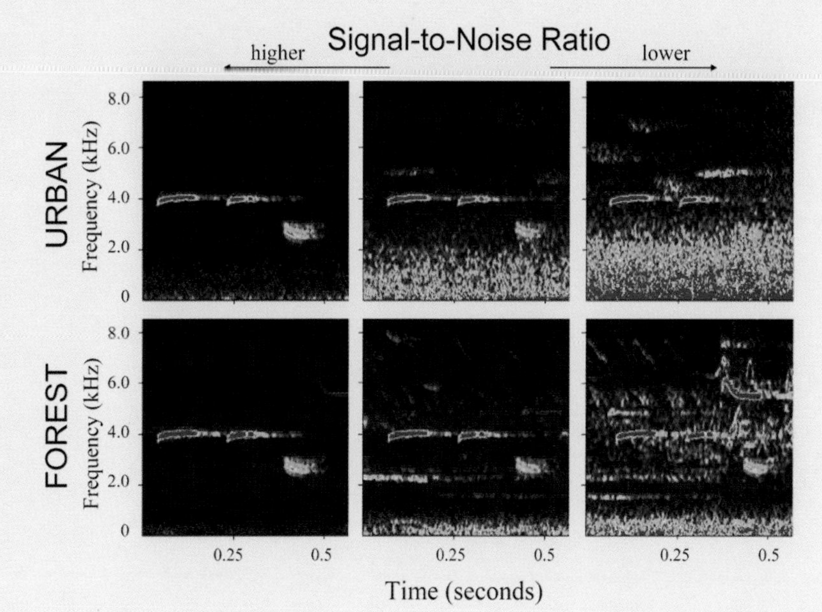

Evolution of the chattering classes

The complexity of human language is one of the factors that has led many people to regard us as a breed apart from the rest of the animal world. But like many types of behaviour that once seemed distinct and are now regarded as shared attributes, research is chipping away at the notion that humans are the only talking species.

In Nigeria's Gashaka Gumti National Park, a team of scientists dis-covered the first evidence that another species may be able to grasp rudimentary syntax and string a sentence together. They observed the putty-nosed monkey combining individual sounds to give them a different meaning to when they are uttered independently.

The "pyow" sound on its own is used by the monkeys to warn of the presence of a leopard, while a cough-like noise described as a hack indicates an eagle is hovering nearby. Used together by a single male accompanying a group of females and young, a string of up to three pyows and four hacks translates as, "Let's get out of here". It is a signal that is used both when predators are lurking and when the male decides it is simply time to move on from the area to find somewhere else to forage or sleep.

The researchers from the University of St Andrews in the UK concluded that the calls were produced intentionally rather than randomly. Dr Kate Arnold said the monkeys provided the first good example of animal communication where call combinations had different meanings to the constituent elements. This, she said, suggests the ability to form basic sentences may be a trait found in other primates.

Previous research undertaken by Dr Arnold's colleague, Dr Klaus Zuberbühler, showed that putty-nosed monkeys moving into areas dominated by Diana monkeys were tolerated, at least when food was abundant, because their alarm calls served both species.

A string of up to three pyows and four hacks translates as, "Let's get out of here"

right: Putty-nosed monkey pictured in Nigeria's Gashaka Gumti National Park [photo courtesy of Kate Arnold]
opposite: Putty-nosed monkey native to West African rainforests [photo: Mark Bowler / NHPA]

Shopping in the wild

Several research projects have been assessing how far, if at all, animals look to the future by planning their actions. Until these were carried out, it was unclear whether any species other than humans are able to actively think ahead rather than simply react to seasonal cues and other external stimuli.

Squirrels, for example, have long been observed to hoard nuts to give them a supply of food over winter. But it is uncertain if they know winter is coming and therefore decide to guard against starvation, or if they are acting instinctively because they are programmed to hide surplus food when autumn begins.

Researchers are now beginning to show that, at least for some animals, there is an element of conscious foresight involved in their behaviour, revealing a greater complexity of mental capacity than previously accepted.

Baboons in the Blouberg Nature Reserve in South Africa, for instance, have been found to go on shopping expeditions, behaving much in the way that human bargain hunters do when shops hold sales. According to a study conducted by researchers from the University of St Andrews in the UK, the baboons know that if they are slow to head to a particular tree after waking up in the morning, other animals will beat them to all the tastiest fruit.

The baboons' response is to think ahead—once awake they head directly to that tree and ignore other sources of food on the way, thus ensuring that they get to the fruit before it has all been grabbed. Having done this they can have a more relaxing day, searching out other trees and bushes for the less appetising or more commonly available foods that are not at risk of disappearing early in the day.

opposite: Yellow Baboon male, *Papio cynocephalus*, in Amboseli National Park, Kenya [photo: Martin Harvey / NHPA]

The study found that baboons are able to imagine what is out of sight, decide which destination has priority, and plan accordingly—just as human shoppers do

Professor Richard Byrne, who took part in the study, said the findings show that baboons are able to imagine what is out of sight, decide which destination has priority, and plan accordingly—just as human shoppers do. He maintained that the ability to plan ahead and display foresight is likely to go back some 30 million years to the joint ancestor of mankind and baboons.

Similarly, scientists in Germany, at the Max Plank Institute in Leipzig, have established that all the great apes are likely to have the facility of foresight. They looked at captive orangutans and bonobos, a species very similar to chimpanzees, and found that each of them chose tools suitable for particular food-related tasks well before using them.

With the ability to plan being displayed by humans, orangutans and bonobos, it seems likely that it is a facility also present in chimpanzees and gorillas, which are part of the same evolutionary group that descended from a common ancestor dating back about 14 million years.

Where's my breakfast?

A thieving bird capable of "mental time travel"

above: Western scrub-jay [photo: Kevin T. Karlson]

Perhaps more surprising than baboons planning shopping trips is the discovery that foresight can be found in animals that are from species outside the primate groups. Western scrub-jays, a North American bird species, have been the focus of work by the UK's University of Cambridge for several years.

To assess whether the birds were capable of planning ahead, they were given breakfast only on alternate days. One day they would get breakfast, and the next would go hungry until later in the day. The birds had access to different compartments where breakfast would either be served or withheld. Overnight the bird would not be able to get into the compartment, but could during the day.

After six days, the birds demonstrated the ability to forecast a potential future problem, as they began to put aside nuts from other meals during the day and hoard them in the compartment where they realised breakfast would be withheld the next morning. They put aside, on average, 16.3 pine nuts for the following day's breakfast, which ensured something to eat when they woke up.

A second experiment, where breakfast was provided daily but alternated between dog food and peanuts, confirmed the birds were able to envisage what the future held. Once again, they quickly realised the first meal of the day could be improved, this time by variety. On days when peanuts were offered, the birds stored some in the compartment where they knew they would find dog food the following day. Similarly, they placed dog food in the spot where peanuts would turn up at the next breakfast.

Professor Nicky Clayton said the study clearly demonstrated the ability of the scrub-jays to plan—the first time it has been shown in a non-primate species, and the first time any animal was observed to look ahead to what tomorrow would bring.

The study builds on earlier work by the experimental psychology team, which showed that the bird is capable of "mental time travel", or in this case, that it takes a thief to know a thief. Scrub-jays cache food for later when given the opportunity, and are not above raiding another bird's food store if they saw it being hidden. Observations backed up by laboratory experiments revealed that the scrub-jays with experience of stealing from other birds will know if their own caches are at risk. Having hidden the food, they will return and re-hide it if they knew they were being watched by a rival during the initial caching. However, those scrub-jays that had not stooped to thievery themselves entertained no suspicions of other birds, so made no attempt to re-store hidden food.

Professor Clayton said experimental proof of the birds' ability to project their own experiences to assess the possible motives of another scrub-jay was a major development in the study of animal cognition and the first time it had been demonstrated in non-primate species. The ability to calculate another's intentions, desires and beliefs develops in humans at about the age of three.

Snail trails

Efficiency is important to animals. When the next meal is rarely guaranteed, it is essential to waste as little energy as possible. Even snails, hardly the fastest-moving creatures, will conserve energy when they can.

Snails and slugs leave slime trails behind them as they move and in the past it was suggested that these are left as signposts pointing the way towards food sources. However, it has now been shown that reusing the trails is an effective means of saving energy. By hitching a ride on ready-laid mucus trails, the common periwinkle snail, *Littorina littorea*, expends as little as 27 per cent of the energy that it requires to create an entirely fresh track.

The finding, expected to apply to other snails and slugs, was made by researchers from the University of Sunderland in the UK, who measured the thickness of the trails left by the marine snail. The mechanism by which the snails read the thickness of an existing trail and calculate how much slime they need to lay themselves has yet to be established. But with a third of what snails consume estimated to be needed to fuel mucus production, there are clear benefits to be gained—lots of left-over energy to find and eat all the flowers in gardens.

Reusing the slime trails is an effective means of saving energy

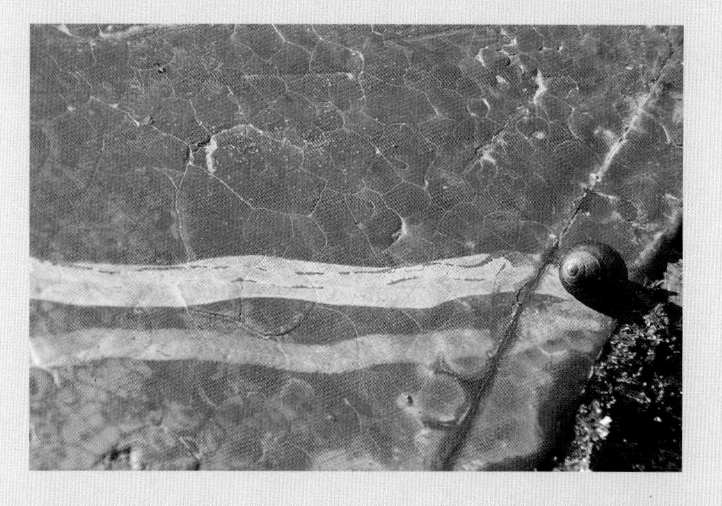

right: Common periwinkles and their trails
[photos courtesy of Mark Davies, University of Sunderland]

Clever fish

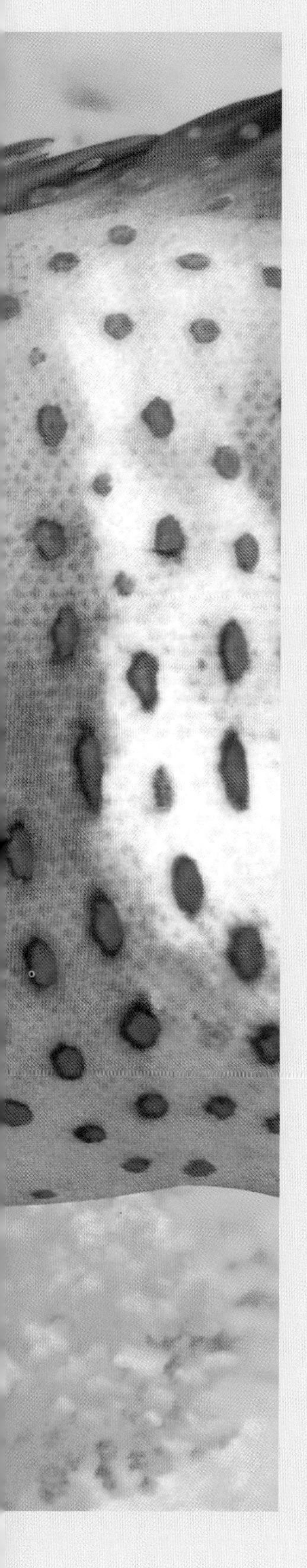

By hunting as a team, the eel and the grouper give prey species nowhere to hide

Cooperation between different species has been discovered between groupers and giant moray eels in the Red Sea.

The two types of fish join forces to go hunting and swim side by side, as if taking a stroll together, according to researchers at the University of Neuchatel in Switzerland. They make an ideal predatory team—the eel can get into the nooks and crevices of reefs, while the grouper specialises in catching prey in open water. By hunting as a team, prey species have nowhere to hide and are driven into the path of one of the companion predators.

Discoveries of such behaviour add to the growing evidence that fish are not as dumb as they look. Rainbow trout were the subjects of one study of fish behaviour that demonstrated that they have different personalities and quickly learn from their experiences. It is the bolder trout that is most likely to end up on a hook because it is less cautious about what it eats. Furthermore, it needs to eat more since it uses up more energy through being energetically curious. But the trout can learn to become shy and retiring, thus protecting it from the baited hook, if it goes through a nasty experience.

Experiments led by Dr Lynne Sneddon of the University of Liverpool in the UK demonstrated the trout's ability to learn. The fish were initially categorised as shy or bold, depending how long they took to investigate a small toy building brick dropped into their tank. The next stage was to provoke a fight between the aggressively territorial trout by placing two fish, one bigger than the other, in a tank together and recording the outcome. Shy fish that repeatedly won the fights became noticeably bolder, while bold fish that repeatedly lost became more timid, revealing an ability to learn from experience.

Unsurprisingly, bold fish that won their bouts became even more adventurous but, less predictably, shy fish that lost again and again also became more adventurous, rather than becoming even more reclusive. The research team suggested a "desperado effect" came into play.

left: Grouper, *Plectropomus pessuliferus* [photo courtesy of Redouan Bshary, University of Neuchatel]

Deadly water pistols

Archer fish have been shown by a team at the University of Erlangen in Germany to be even more impressive at squirting jets of water at prey than previously realised. It is not a question of fire and hope—the fish are deadly accurate and can calculate the quantity of water needed to dislodge each unfortunate creature chosen for supper.

The quantity of water spat at their prey—generally flies, beetles and spiders, and occasionally small lizards—is judged according to size. Researchers found that the fish consistently squirted about ten times as much as was needed to knock the chosen prey into the water where it could be snapped up.

The archer fish's hunting technique is costly in terms of energy expended, so it limits wasted effort by altering the amount of water expelled according to the creature it intends to dislodge from an overhanging plant. Of course, using ten times as much water as is required would seem to be wasteful in itself, but it is thought to be a built-in safety margin to maximise the chances of success.

The findings came just two years after it was demonstrated that archer fish can assess not just the size and position of their target, but can also take account of distortion caused by light striking the water.

above: Archer fish [photo courtesy of V. Runkel and S. Schuster, University of Erlangen-Nurnberg]
opposite: Archer fish firing jet of water from mouth to bring down prey [photo: Stephen Dalton / NHPA]

The fish are deadly accurate and can calculate the quantity of water needed to dislodge whichever unfortunate creature they've chosen for supper

Spy in the sea

Sensors of one form or another have proved a boon to researchers trying to study animals, whether insects, big cats or marine creatures. They can be simple tracking devices that enable scientists to find them at will, or tags that record a variety of behavioural data. They go where the animals go, which people often cannot do, or, if they did, would interfere with the subject's behaviour.

Even for fairly well-understood species, tags can prove useful. Brent geese tagged as part of a project involving the Wildlife and Wetlands Trust in Slimbridge, UK, were tracked on their migration routes and, among other details, the devices revealed a previously unknown breeding area on a Norwegian island in the Svalbard archipelago in the Arctic.

On animals like penguins, tagging devices can provide a wealth of information that is beyond the scope of research that relies on visual observation. Professor Rory Wilson of the University of Wales in Swansea, UK, has developed several different tagging devices since the early 1980s to facilitate the research of marine life.

One of the most recent discoveries the tags have established is that Magellanic penguins are among the most voracious creatures in the sea, eating the equivalent of a grown man consuming almost 600 quarter-pound burgers in eight hours.

Because the tags record the muscle movements associated with catching and swallowing fish or squid, the devices were able to show just how much the penguins ate under the waves. Until the recording tags were developed, the only way to measure the penguins' appetites was to catch them and view the stomach contents after giving them an emetic.

The tags showed that Magellanic penguins can eat up to three times as much as previously thought. This is, of course, unlikely to make fishermen love them, but does provide important information that can be used by conservationists.

Further details of the fishing techniques of Magellanic penguins were revealed by the devices, which record data such as heart rate, depth, location, direction, speed and temperature. As had been suspected but never proved, they herd fish like sheepdogs do sheep.

The penguins herd fish like sheepdogs do sheep. Swimming round in ever decreasing circles, the penguins force their prey into balls that eventually break up

Swimming round in ever-decreasing circles, the penguins force their prey into balls that eventually break up. The penguins can then snatch up the stray fish.

With the devices able to detect the number of breaths penguins take before diving, Professor Wilson was able to show that they calculate how deep they are going to dive and how many fish they anticipate catching.

opposite and overleaf: Magellanic penguins
[photos courtesy of Rory Wilson, University of Wales, Swansea]
below: Tagged brent geese
[photo courtesy of WWF / Kendrew Colhoun]

Furthest, fastest, deepest

Tags used on Cuvier's beaked whales tracked a record-breaking dive to 6,230 feet (1,899 m) below the surface of the sea, in which the whale held its breath for 85 minutes. But unlike sperm whales and elephant seals, which can make repeated deep dives, the beaked whale appeared to need a rest in shallow water after trekking to the depths, probably to recover its oxygen levels. The measurement was taken during research by scientists from Woods Hole Oceanographic Institution in the US, who were investigating the suspected link between whale strandings and the use of sonar.

Tags recording position, air temperature and depth of diving allowed the sooty shearwater to claim the crown for the longest migration. The bird, which only weighs about two pounds, flies up to 46,000 miles (74,000 km)—more than twice the 22,000 miles (35,000 km) flown by the Arctic tern, which was previously recognised as having the longest migration.

To carry out the study, researchers caught 33 birds at breeding burrows in New Zealand and tagged them. The following year, 20 of the tags were recovered after the birds had flown around the Pacific Ocean in a figure-of-eight pattern. Further tag data showed the sooty shearwaters dived up to 225 feet (69 m) into the sea to catch squid, krill and fish, while the average dive was 46 feet (14 m).

High-speed videography was needed to measure the speed of a trap-jaw ant's bite, which at up to 145 miles per hour (233 km/h) turned out the be the fastest predatory strike in the animal kingdom. The jaws are spring-loaded and snap shut in an average of 0.13 milliseconds with an acceleration 100,000 times the force of gravity. It all takes place 2,300 times faster than a human can blink an eye, according to calculations by a team led by Dr Sheila Patek of the University of California, Berkeley, in the US.

above: Sooty shearwater from Whenua Hou Island, New Zealand [photo courtesy of Josh Adams, © 2005]
opposite: A small cricket falls prey to a trap-jaw ant [photo: Alex Wild, www.myrmecos.net]

The Central and South American ant, *Odontomachus bauri*, also uses its jaws to escape when in danger. When threatened, the ant will release its jaws either against the predator or the ground, catapulting itself up to 15.6 inches (39.6 cm) away—the equivalent of a 5'6" (1.68 m) human leaping 132 feet (40 m).

The ant's jaws are spring-loaded, snapping shut with an acceleration 100,000 times the force of gravity

When threatened, the ant can catapult itself with its jaw more than 15 inches away—the equivalent of a 5'6" human leaping 132 feet

right: Trap-jaw ant [photo: Alex Wild, www.myrmecos.net]
below: A series of stills from a video showing how the trap-jaw ants, *Odontomachus bauri*, fire their mandibles with such acceleration that the forces generated are strong enough to propel the ants' bodies through the air. In the "escape jump", shown here, the ant's trajectory is directed upward [video courtesy of Dr Sheila Patek and colleagues, University of California, Berkeley]

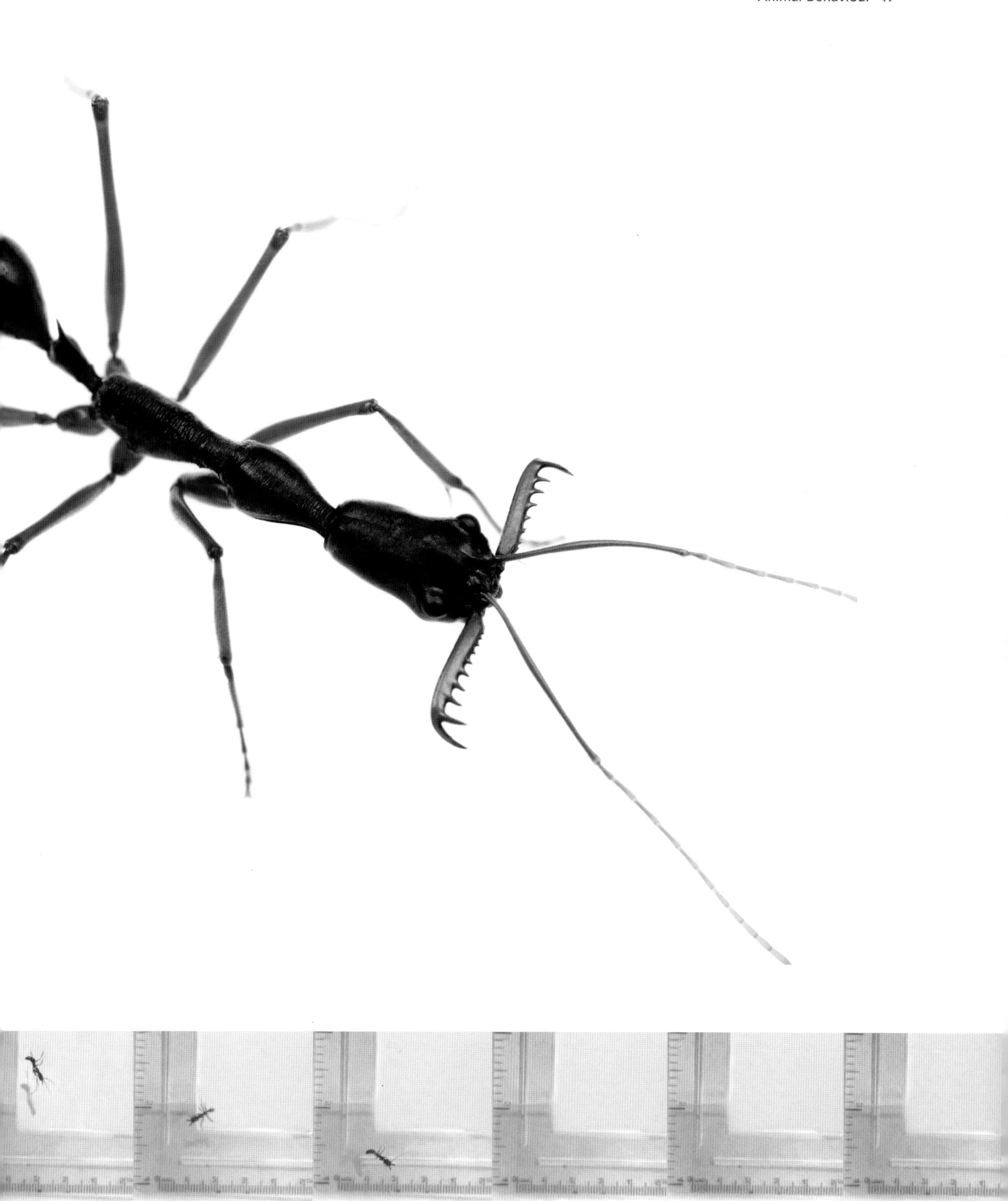

Darwin in action

One of the most startling examples of evolution in action has been found among the very finches of the Galapagos Islands that helped make the name of Charles Darwin and his survival of the fittest theory.

The medium-sized ground finch, *Geospiza fortis*, found itself in competition from 1982 with the introduction of a bigger finch, *G. magnirostris*. The bigger finch was better able to break open the seeds of the *Tribulus cistoides* plant, which was until then the smaller ground finch's main food source. This began to push the smaller finch towards a smaller beak size better suited to get at the smaller seeds of other plants, according to research led by Princeton University in the US.

In 2004 a drought led to a shortage of the *Tribulus cistoides* seeds and most of the bigger finches were wiped out. Similarly, the smaller finches that had retained comparatively large beaks starved. The survivors were predominantly those finches that had developed smaller beaks. Measurements taken after the drought revealed that the average beak size of *G. fortis* had fallen dramatically despite the body size remaining the same, showing that the bird had evolved to cope with survival pressures of competition and seed shortage.

In the space of only a year, the brown anole lizard's leg length evolved twice to avoid a predator

Even faster to evolve under pressure was the brown anole lizard, *Anolis sagrei*, during experiments on tiny islands, or cays, in the Bahamas. In the space of only a year the leg length of the brown anole evolved twice to avoid a predator, the northern curly-tailed lizard, *Leiocephalus carinatus*, which was introduced into its habitat. In the first six months the anoles with longer legs were most likely to survive, as they were better able to outrun the predator.

Over the second six months the legs became shorter thanks to a behavioural change. Instead of spending most of their time on the ground, the anoles became more arboreal, clinging to thin branches and stems where they were in less danger from the predator lizards. This behavioural change, documented in research led by Professor Jonathan B. Losos of Harvard University, US, meant that by the end of just 12 months the surviving anoles had much shorter and stumpier legs.

opposite: Medium ground finch, Galapagos Islands [photo: David Middleton / NHPA]
right: Brown anole lizard [photo: Melvin Grey / NHPA]

Batting for koalas

Artificial insemination techniques have been developed to help protect the koala from inbreeding and sexually transmitted diseases. The programme has proved so successful that pregnancy rates from artificial insemination almost match the natural method, and moreover it is a better guarantee of a healthy population.

The Australian marsupial's population numbers have slumped from several million a century ago to an estimated 40,000 to 100,000 today, largely due to the impact of humans. Falling fertility rates because of sexually transmitted diseases have compounded the problem, as has inbreeding, a result of koalas being penned into areas of habitat surrounded by roads and buildings. Researchers from the University of Queensland in Australia, aided by the Zoological Society of London in the UK, believe the artificial insemination project offers a safeguard that will help ensure the animal's long term future.

Collecting the sperm from the males has been made easier by the marsupial's single-minded approach to mating. Once the male has mounted the female, nothing distracts him—not even a scientist nipping in with a test tube. The males are completely oblivious to the careful positioning of the tube, which is lined with rubber from a cricket bat handle for a comfortable fit, and carry on regardless.

The Australian marsupial's population has slumped from several million a century ago to no more than 100,000 today

left: Koala female, *Phascolarctos cinereus*, with young on back [photo: A.N.T. Photo Library / NHPA]
overleaf: Koala female and young in the eucalyptus forests of Eastern Australia [photo: Dave Watts / NHPA]

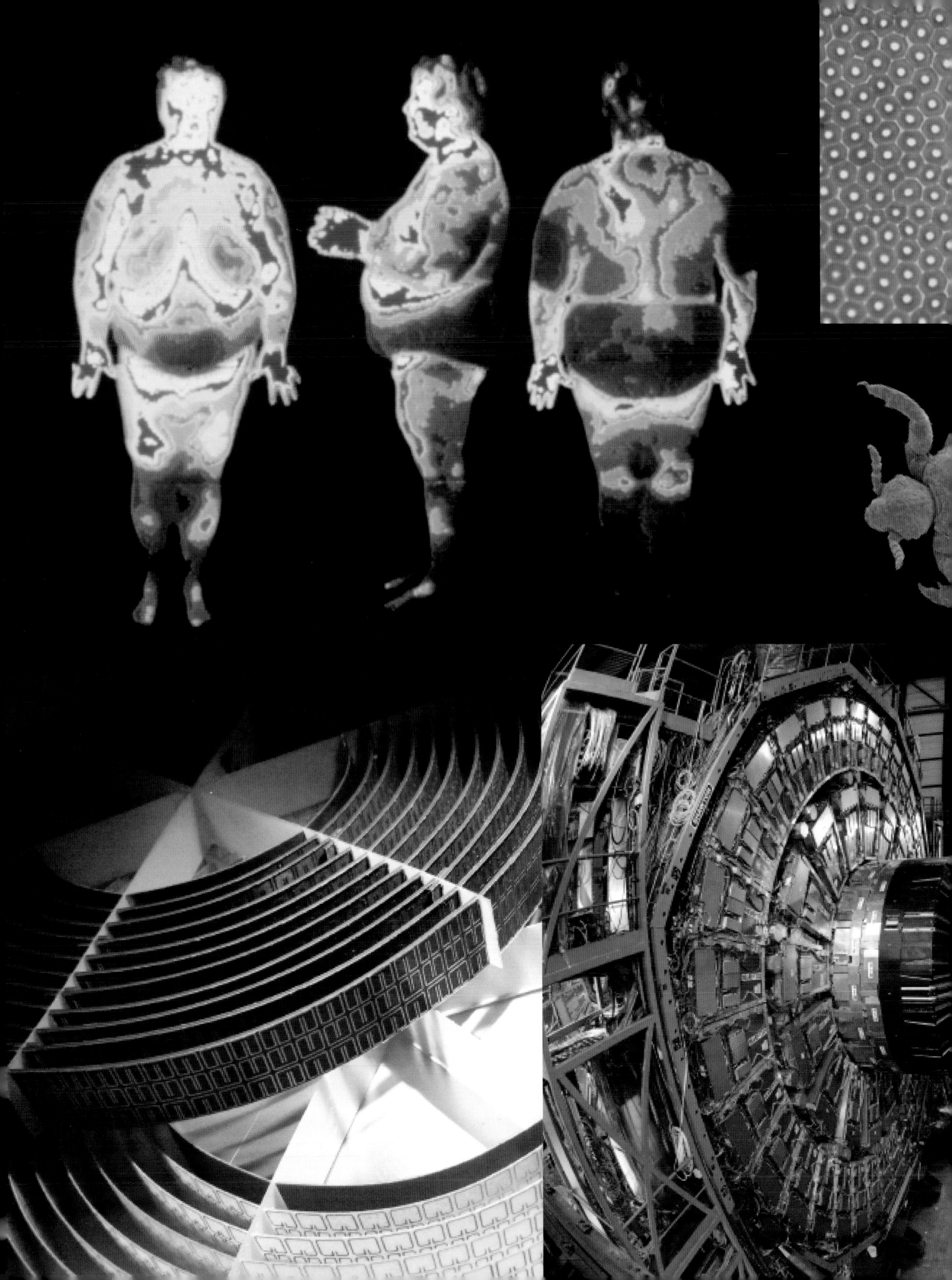

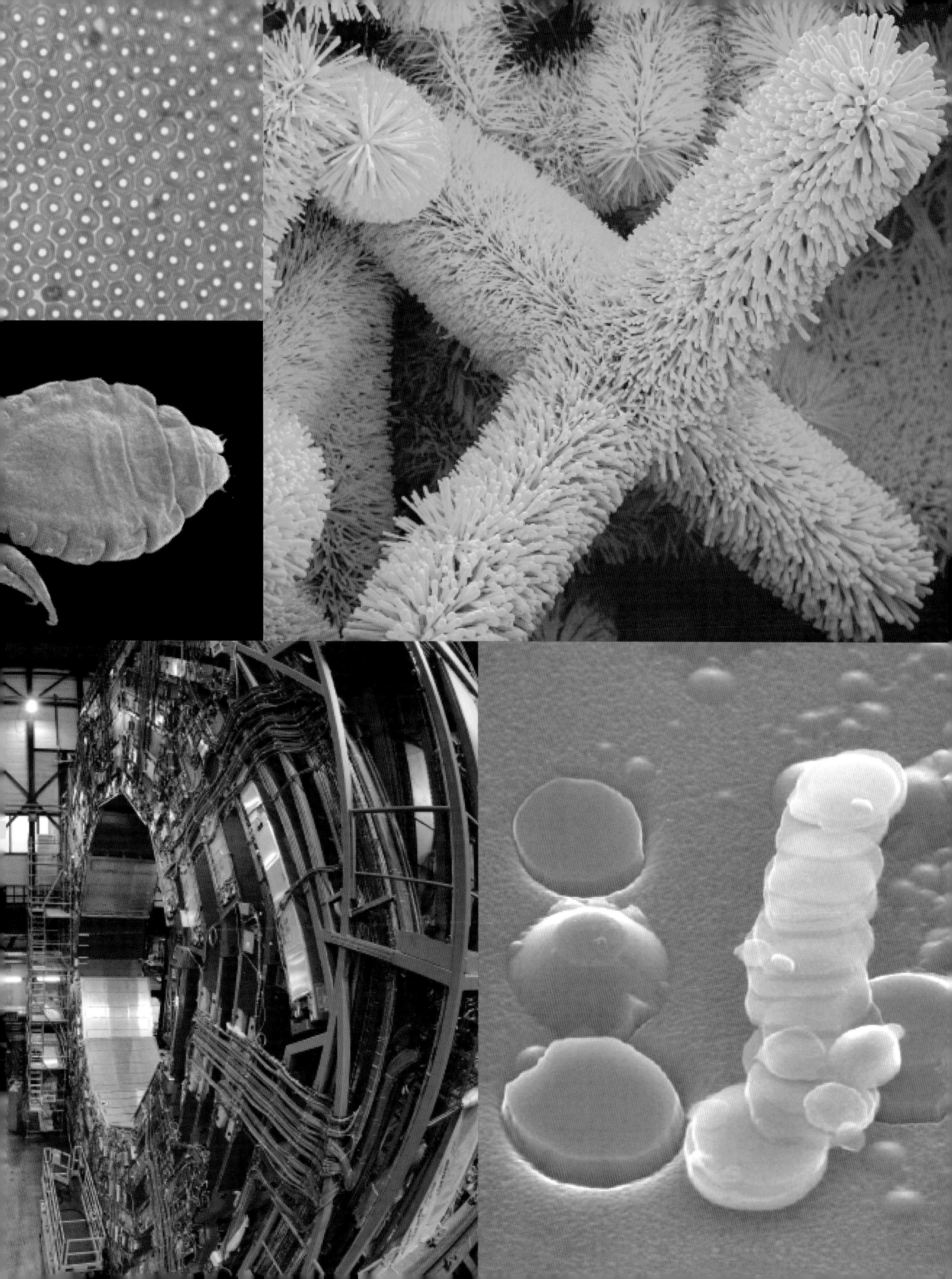

Previous, clockwise from top left: Thermograms of an obese woman [© Dr Ray Clark & Mervyn Goff / Science Photo Library], image detailing the *Plusiotis boucardi* beetle's honeycomb pattern on the central region of its back [courtesy of Dr S. A. Jewell, University of Exeter], zinc oxide nanostructures [image courtesy of Zhong Lin Wang, Georgia Institute of Technology], graphene nanofabric [image courtesy of Andre Geim, Mesoscopic Physics Group, University of Manchester], view of CERN's CMS detector in the surface hall at Cessy [photo courtesy of Maximilien Brice; © CERN], section of invisibility cloak [image courtesy David R. Smith, Duke University], head louse [image courtesy of Vincent S. Smith, Natural History Museum, London]

Tomorrow's World

The pace of new technology has swept along at breakneck speed in the last 50 years, revolutionising our culture in the process. A child born in Western society half a century ago has lived to see an era of unparalleled luxury, where cars are ubiquitous, travel to the other side of the planet is commonplace and food is available in such abundance that obesity has become a widespread health problem.

Robotic technology, computers, a global communications network and inexpensive worldwide transportation have transformed industry. Cheap televisions, telephones, computers, refrigerators, central heating, and labour-saving gadgets such as vacuum cleaners, washing machines and dishwashers have changed the home beyond our grandparents' recognition and are now taken for granted.

On the medical front the contraceptive pill ushered in a social and sexual revolution, cancer is no longer an automatic death sentence and the first test-tube babies are now producing their own offspring. Just a generation ago technological advances such as the mobile phone, internet, virtual reality, genetic screening and a host of infertility treatments seemed the stuff of science fiction, only to become commonplace today. Invisibility cloaks, bionic eyes, malaria-busting mosquitoes, robot doctors and drivers, and fat-burning pills may seem even more fantastical, but, according to today's researchers, could be just around the corner. The pace of change remains relentless.

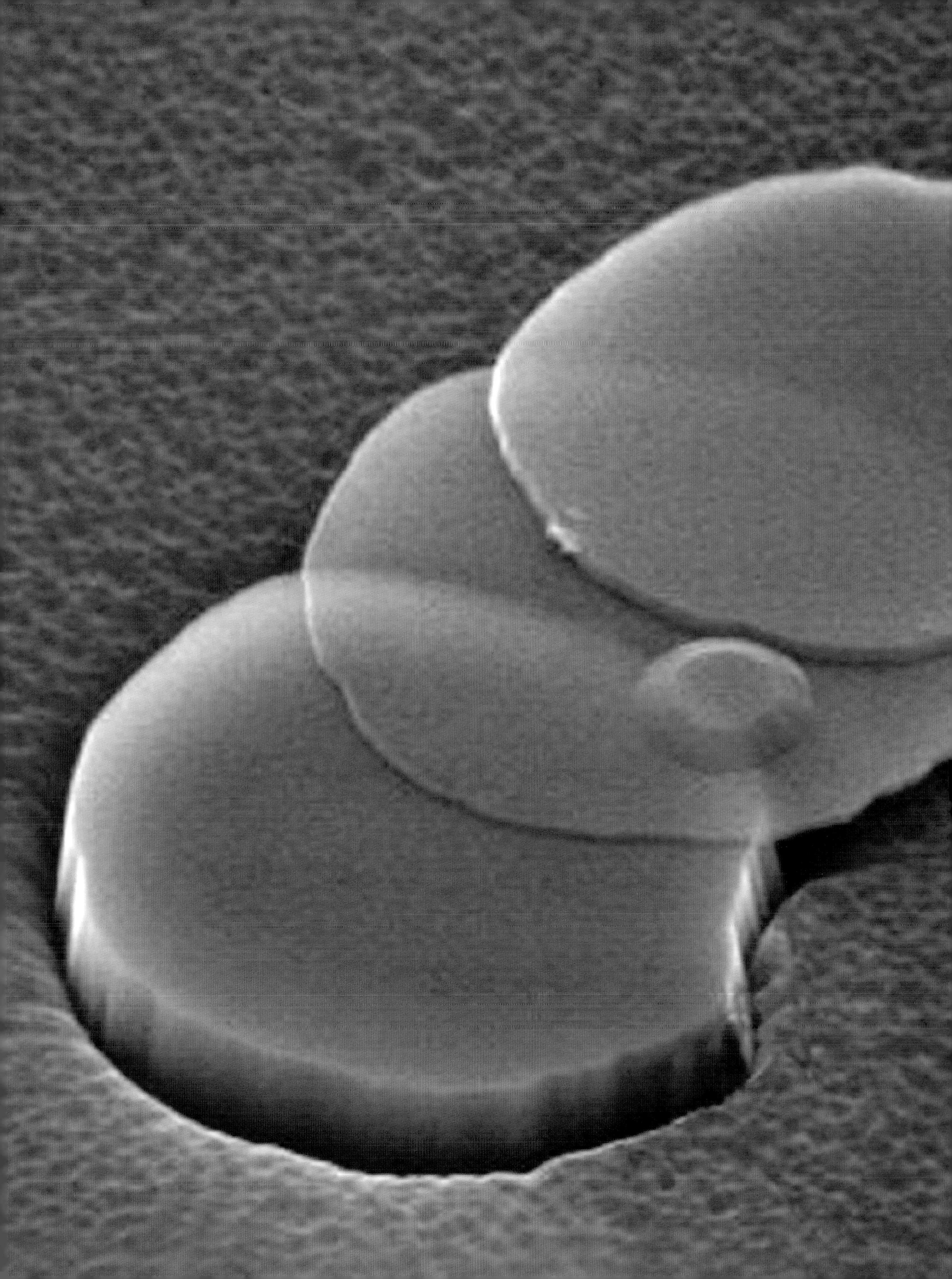

Thinner than size zero

Few technological advances have matched the impact of silicon in computer chips, but the demand for ever faster and more efficient data processing means that the search for its successor is well underway. Graphene, a material made from carbon molecules, is one of the most promising developments by researchers looking for a successor to silicon's mantle.

About 200,000 times thinner than a human hair, graphene is the thinnest material ever created. Just one atom thick, it is regarded as a two-dimensional substance and has properties that allow electrons to travel along it at high speed—giving it the potential to operate as an ultra-fast transistor.

Graphene originates from graphite, which is constructed of a series of two-dimensional layers of atoms. In experiments led by scientists from the University of Manchester in Britain and in Chernogolovka, Russia, it was proved possible to create graphene by cutting away ultra-thin slices of graphite using a technique called micromechanical cleavage.

According to past theory, crystalline materials one atom thick, regarded by science as two-dimensional, were an impossibility, as the slightest thermal change would destroy them. Experiments on thin films first backed up the theory when it was shown they were unstable below a certain thickness. Graphene, however, defied the theory when researchers managed to cleave layers just one atom in thickness that remained structurally stable even at room temperature—a crucial requirement of any material operating in a microprocessor.

But graphene as it was originally created could only exist when stuck to another material, and its early promise as a replacement for silicon in computer chips ebbed away as follow-up experiments suggested it was far less efficient than first predicted.

In early 2007, however, it was revealed that the Manchester team, this time working in partnership with the Max Planck Institute in Stuttgart, Germany, had succeeded in creating a film of graphene that was free of backing material. It was once again a front-runner for the chief ingredient of tomorrow's computer chips. Electron refraction analysis of the membrane hanging from microscopically small gold scaffolding showed that the one-atom layer undulated. The wavy structure is thought to be the reason why it has enough strength to remain stable despite being so thin. Further experiments on its potential as a high-speed transistor, to act as a switch to allow in or keep out electric current, suggest there is every chance it can be developed to take over silicon.

Applications can also be expected in other fields, not least in medical research where graphene could be used in electron microscopes. Graphene is so thin that its chicken-wire structure has the potential to be used to sieve gases. That same structure could support proteins and other substances that medical researchers want to analyse under electron microscopes. Because it is so thin, the resolution of the readings from the microscopes could be much higher since the electrons would have to travel through far less extraneous material.

At 200,000 times thinner than a human hair, graphene is the thinnest material ever created

right and below: Scanning electron micrographs of a fallen mesa of graphite, from which graphene molecules were "extracted" [images courtesy of Andre Geim, Mesoscopic Physics Group, University of Manchester]

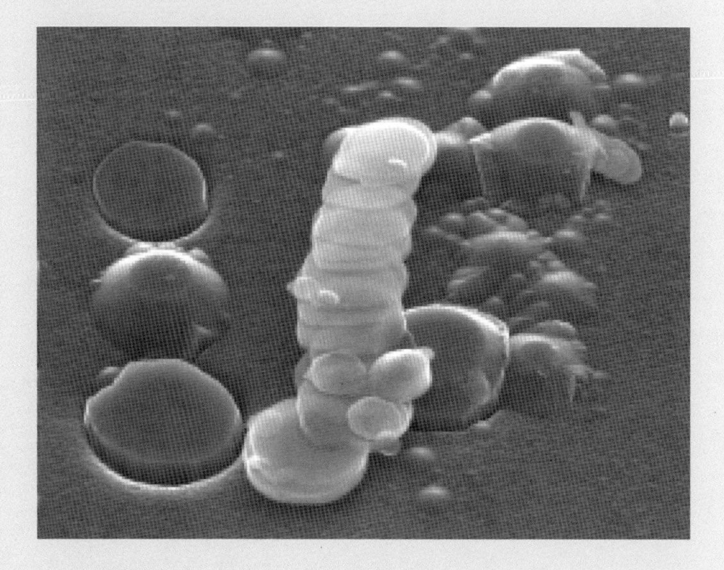

Invisibility cloak

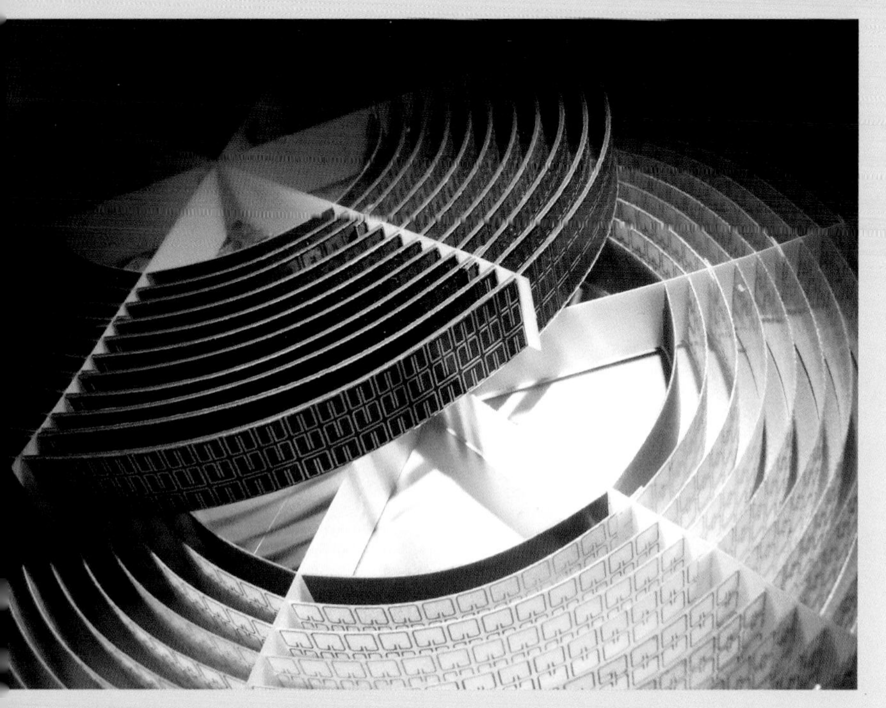

left: Microwave cloaking structure
below: Snapshots of electric field patterns, with the black stream lines indicating the direction of power flow [images: Science / AAAS]

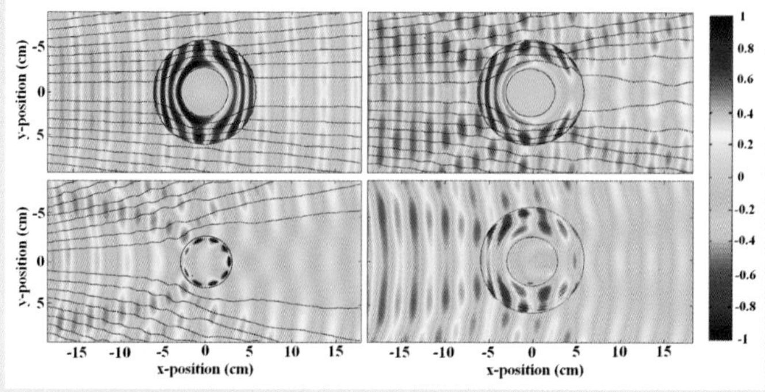

Just as graphene seems to exist in defiance of the laws of physics, so the idea of invisibility seems to defy common sense. Nevertheless, research suggests that far from being limited to works of fiction such as *The Invisible Man* and *Harry Potter*, invisibility could become a very real fact of life.

Researchers at Duke University in North Carolina, US, have managed to make a small object disappear, at least when microwave beam sensors were employed. The invisibility device was constructed of concentric rings of copper and metamaterials, a type of artificial composite, which deflected microwave radiation around an object placed in the centre. Much as water flows past a rock in the middle of a stream, the microwave beams flowed around the object in the middle and rejoined once past, giving the appearance to external sensors that they had continued unimpeded. At this stage the device, at five inches (13 cm) in diameter, is small and the hidden object even smaller, but the technology has the potential to be expanded dramatically.

The US success in creating an invisibility shield drew on the work of Professor Sir John Pendry of Imperial College London in the UK, who proposed the design theory and electromagnetic specifications. His theories were dependent on the creation of new materials, the metamaterials, which react differently to natural materials when exposed to electromagnetic waves. He doubts the technology will be practical for concealing large buildings or light enough to hide aircraft, but is confident it will be developed sufficiently to make tanks invisible, perhaps as soon as 2011.

Of course, the deflection of microwaves is a long way off from warping light waves to make objects vanish before our eyes. It is, however, the first step in that direction and it is not inconceivable that in time the same principles will be applied to visible light, possibly giving rise to the ultimate in merchandising—a working Harry Potter invisibility cloak.

> Rather than remaining confined to works of fiction like *Harry Potter*, invisibility could become a very real fact of life

Sugar in the tank

On a more mundane but arguably more useful level, research into everyday batteries suggests the next generation could be powered by fizzy drinks or plant sap. A battery under development at Saint Louis University in Missouri, US, has successfully kept a portable calculator going using a range of sugary drinks as fuel. Sugary liquids are poured into the battery and enzymes break down the sugar into electrons and protons to provide electricity, with water the main by-product.

Sugar water, still soft drinks, glucose and tree sap have all been tested on the prototype and shown to be useful sources of electricity. Fizzy drinks were also used, but it was found they were much more effective when they had been left to go flat before being put into the battery. The research, led by Dr Shelley Minteer, has been funded by the US military, which is interested in being able to recharge batteries on the battlefield and in other situations where plug sockets and conventional electricity supplies are absent.

Ideally, pre-filled cartridges would be supplied with the batteries to facilitate recharging whenever necessary. But the utilisation of sugary substances would make it possible for troops to simply reach into their backpacks for a sweet drink should they find themselves running low on power. Equally, being able to squeeze sap from plants in the field of operations would have the potential of being a life-saving convenience.

In the home, such power sources could be equally useful and have the potential to compete with the lithium ion batteries used in mobile phones and laptops, as well as the AA, AAA, D and C batteries commonly used in remote controls, children's toys and torches. While it is not the only sugar-powered battery under development, its creators are confident it is closest to being turned into a commercially viable product, perhaps in less than four years.

Sugar water, still soft drinks, glucose and tree sap have all been shown to be useful sources of electricity

[photo courtesy of Hayley Williams]

Exterminating head lice

A development that could soon find its way into the home is a device designed to end the annoyance and frustration caused to parents and children by head lice.

Lice are regarded with varying degrees of horror. In some places children with head lice are nearly treated as social pariahs, in others an infestation gets little more response than a resigned sigh before a child is tugged upstairs to the bathroom for an extended session of fine-tooth combing and shampooing. Whichever reaction the lice receive, they are universally regarded as a nuisance that takes a surprising level of effort to eliminate—so any quick-fix solution is sure to be welcomed by millions of careworn parents.

Moving away from the chemical treatments most commonly used at present, researchers at the University of Utah in the US have been testing a new system of lice extermination, which is essentially an elaborate hairdryer.

Test results published in late 2006 showed that the device, dubbed the LouseBuster, killed 80 per cent of lice in a single infestation and, equally importantly, 98 per cent of the eggs, or nits, they laid.

Inspiration for the device, which is still under development but could be on the market by the end of the decade, came after Professor Dale Clayton noticed how difficult it was to keep samples of louse infestations in birds alive in Utah. He realised that the region's dry air meant the only way he could keep the lice alive in the laboratory for further study was by keeping them in a humidifier. When his children came home from school with lice, he wondered whether the human head lice form, *Pediculus humanus capitis*, could be similarly exposed to terminal desiccation.

The key to killing the lice and their eggs, he and his colleagues found, was not so much roasting them alive, but drying them out using heat.

The device is similar to conventional hairdryers in that it blows heated air, but has two key differences: the air is cooler than a hairdryer and is pumped out at a faster speed. A specially-designed comb attached to the end of the air nozzle directs the 60°C (140°F) airflow along the scalp in the direction opposite to which the hair is being combed. Repeated combing and blowing leaves the eggs and lice desiccated.

Further testing of the development is required before the device can be put on the market, though it could be ready by the end of the decade for use in schools, nurseries and clinics. The combing and drying technique requires a degree of skill, so it is expected that trained operators—perhaps school nurses who have undertaken a short course—will be needed for the first commercial versions, but eventually the method could be made simple enough for household use.

Repeated combing and blowing leaves the eggs and lice dessicated

above: Head louse [image courtesy of Vincent S. Smith, Natural History Museum, London]
right: Professor Dale Clayton operates the LouseBuster [photo courtesy of Sarah E. Bush]

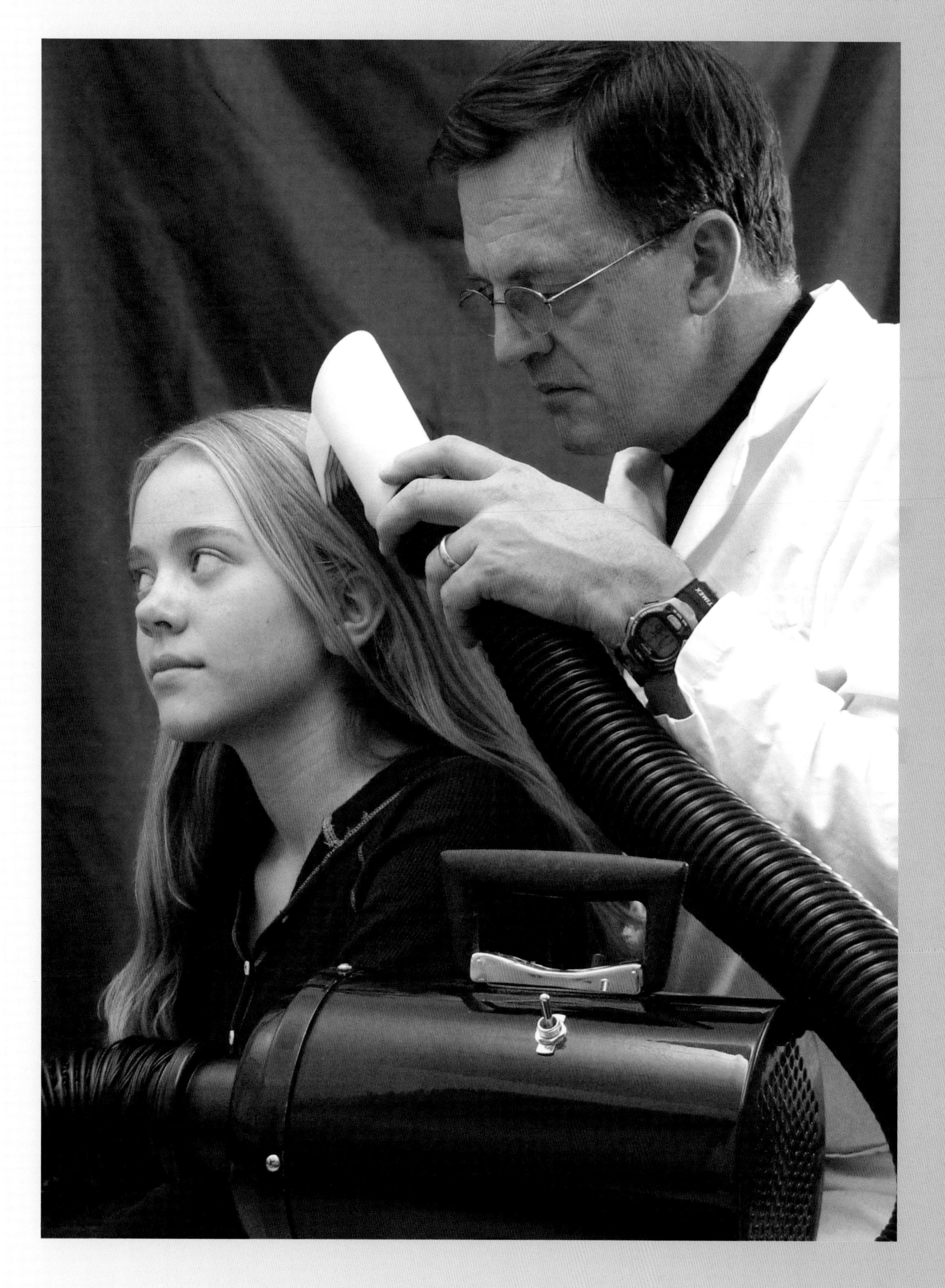

Touchy-feely robots

Robot doctors are a concept beloved of science fiction writers. What better than a doctor who can scan every diagnostic source in a matter of microseconds and carry out open heart surgery with a hand that never shakes?

Robotic surgeons have already established a foothold in the medical world, though as yet only for a handful of simple procedures. Among the many reasons for their extremely narrow use is their limited dexterity, which has in turn been restricted by the absence of an adequately developed sense of touch. Previously, the most advanced robotic dexterity was assessed to be about the same as an average six-year-old child building a house of cards or tying a shoelace. Impressive enough in itself, but falling far short of what is required before letting a scalpel-bearing robot loose on a human patient.

Experiments at the University of Nebraska-Lincoln in the US now point the way to a new generation of robots that will be able to feel with at least the same degree of sensitivity as humans. Researchers created a touch sensor by building up, at nanoparticle scale, a film using layers of gold and cadmium sulphide. A thin sheet of plastic was placed on top of the film, and below it was glass. Placing a coin on the plastic sheet caused changes in the strength of the electric charge and electroluminescence of the nanofilm when a current was passed through the material. The deeper the impression from the coin, the greater the change.

A sensor positioned beneath the glass registered the changes in current and then produced an image revealing that the device was able to feel the shape of the raised pictures on the face of the coin, in this case a US penny. It was accurate enough to be able to show the outline of the face of Abraham Lincoln and the letters "t" and "y" in the word "Liberty", and thus provided a measurement of the pressure being exerted.

A human finger's sensitivity is measured at a resolution of 40 microns, or 40 millionths of a metre. The robotic sensor fell only slightly short of that when the findings were published in 2006, and was about 50 times more sensitive than anything else developed. Since then, the sensitivity has been improved to 20 microns, and the research team has established that the film has similar properties to the sensitivity of the human finger in that it behaves as both a solid and a viscous liquid. This visco-elastic property improves sensitivity when the sensor is in motion, just as it is easier to feel tiny bumps on a surface by running a finger over them rather than just placing it on them. Such sensitivity would have plenty of other uses and it has been suggested that dextrous robots would make ideal astronauts, needing neither to breathe nor to get home once a mission was accomplished. Similarly, they could be used as bomb-disposal experts.

But the ability to hold implements with the same sensitivity as humans would only be one type of use for the sensor. The Nebraska team, led by Professor Ravi Saraf, believes it could be used to provide a quick assessment of whether all the cancer cells in a tumour have been removed by a surgeon. Because cancer tissues are harder than healthy tissues, the sensor might eventually be able to feel a sample and detect clusters of

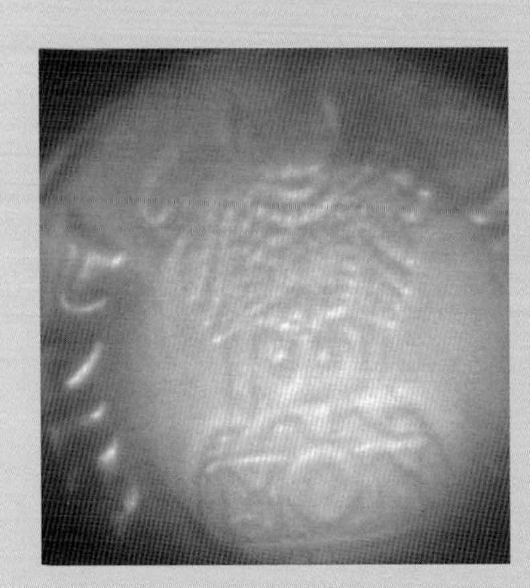

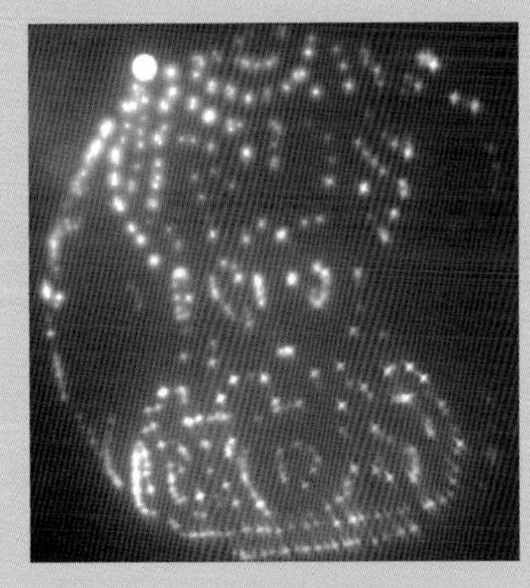

Experiments now point the way to a new generation of robots that will be able to feel with at least the same degree of sensitivity as humans

cancerous cells, informing the medics whether more surgery is required.

Furthermore, the team is developing a robotic touch sensor in hopes that it will be able to detect tumours beneath the skin just by feeling for them. This would help doctors catch cancers earlier in their development because it would be effective at detecting them in lymph nodes, which mammography scans are unable to image. It would also be expected to reduce the number of mistaken readings that lead to unnecessary surgery.

Among the tests the researchers conducted was to place a plastic ball only 0.06 inches (1.6 mm) thick beneath a firm surface. Human fingers were unable to detect the ball, but the robotic sensors provided accurate images.

The future use of robots is seemingly limitless. Robotic lawn mowers and floor cleaners have already been unleashed and it would appear a certainty that as technology advances, so they will become an everyday feature of life. But touch sensitivity is key to robots' success. As Professor Saraf observed, "I personally think that Bill Gates' vision of a robot in every home may not come to fruition if the humanoid cannot do the laundry and fold the clothes."

right: Diagram of a large-area tactile nanodevice for imaging [courtesy of Ravi Saraf]
below: Images of a US penny, where the robotic sensor was able to detect letters in the word "Liberty"
opposite: Robotic sensor images of a coin [images courtesy of Ravi Saraf]

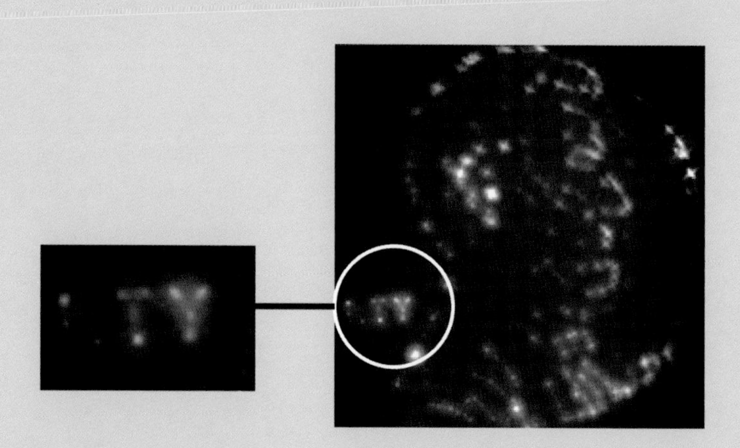

below: Artist's concept of robots for every corner of the home [illustration courtesy of Ravi Saraf, first published in *Scientific American*, January 2007]

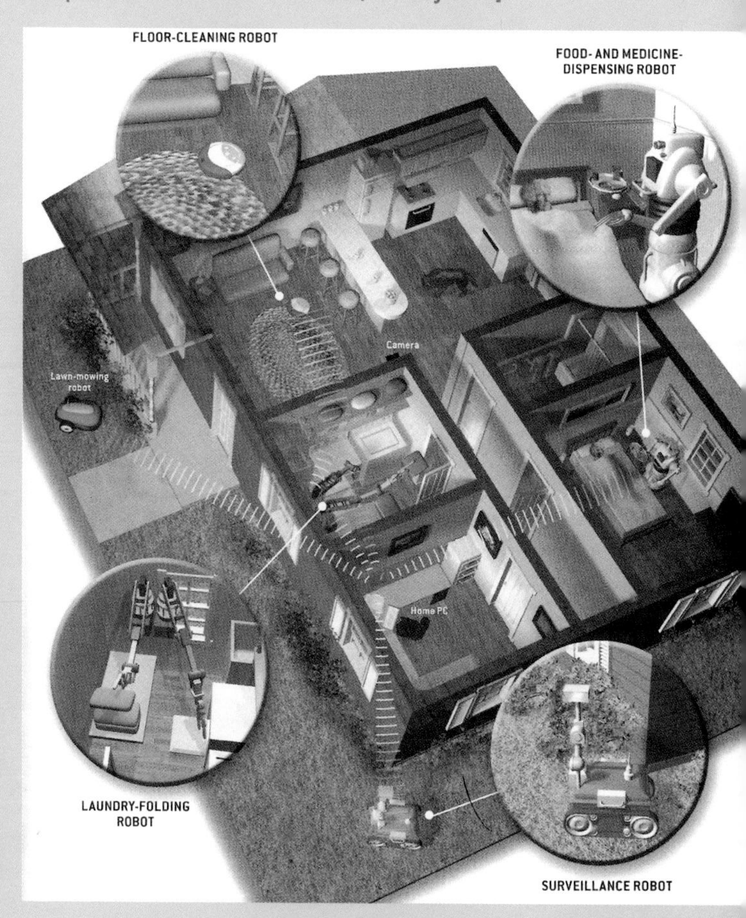

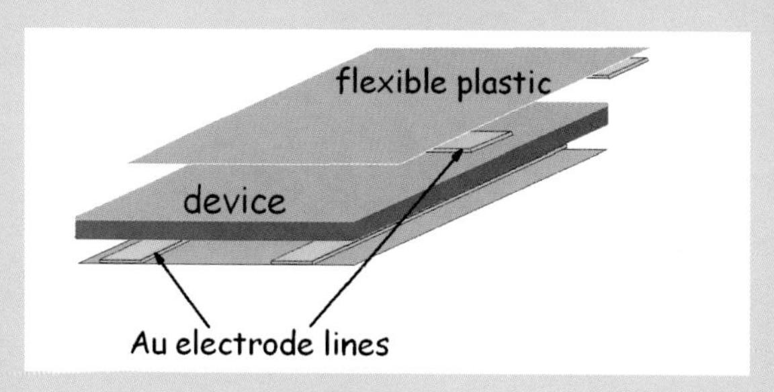

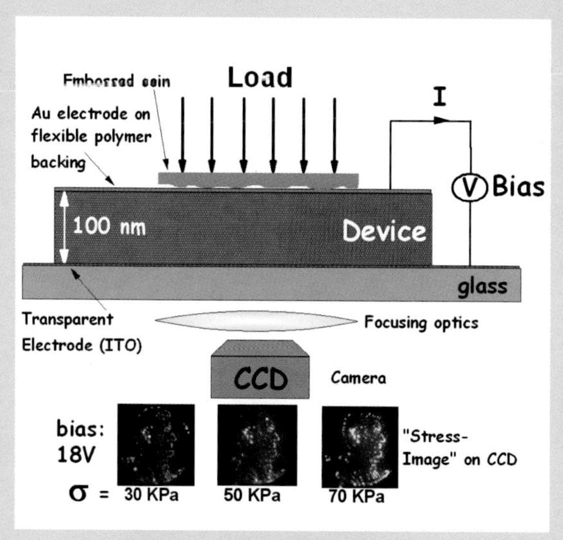

The perfect driver

Driverless driving may seem insane at first thought, but on reflection it has huge potential to cut the number of accidents, big and small. Whereas human drivers have a tendency to lose concentration, focus on one approaching hazard at the expense of another, and even fall asleep, a robot would stay awake the entire journey and be immune to the usual distractions that can lead people to crash their vehicles.

Intelligent sensors are already featured on some cars, such as cameras that detect lines on the road and prevent the vehicle drifting across them, but so far none have threatened to take the wheel from the human driver.

Dr Sebastian Thrun and colleagues at Stanford University in California, US, are among the researchers leading the way towards driverless cars that put robotic systems in control. They won a $2 million prize from a US government agency for successfully modifying a Volkswagen that could detect and avoid stationary obstacles on a 130-mile (209 km) journey. The next step is to avoid other moving cars while observing the rules of the road and signals such as traffic lights.

Military applications are predicted to be in use by 2015, but the real revolution will come in the next 25 years, Dr Thrun anticipates, when cars fitted with artificial intelligence units will begin replacing human drivers, making the road a safer place.

Never sleepy, a robot would stay awake the entire journey

below: Could robot-controlled vehicles prevent this? [photo: iStockphoto / Matt Kunz]

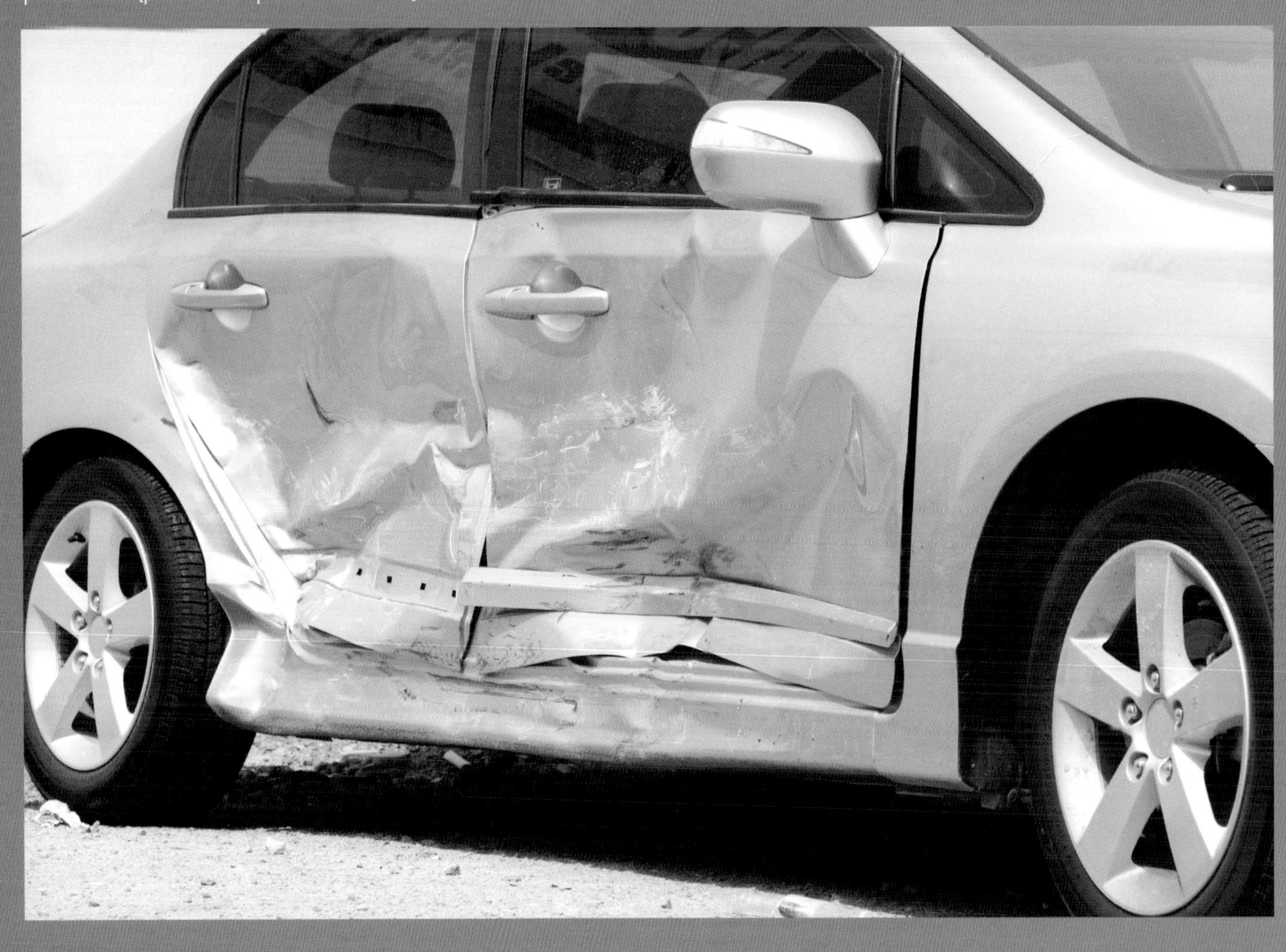

Robo-twitcher

An intelligent camera could save researchers years of fruitless observation based in inhospitable swamp and forest locations

Artificial intelligence has been welcomed by field biologists searching for evidence of the ivory-billed woodpecker. They can now leave much of the donkey work to a device that can recognise the bird in flight. Ivory-billed woodpeckers, dubbed the Elvis bird, were thought to have become extinct in the mid-twentieth century. However, recent unconfirmed, and disputed, sightings in Arkansas, US, have raised hopes it may still survive.

Researchers hoping to verify the sightings realised that an intelligent camera could search for the bird while saving them weeks, months and even years of potentially fruitless observation based in inhospitable swamp and forest locations. And so the world's first robotic twitcher was developed. What's more, it makes no complaint about working through the mosquito and snake seasons in Big Woods, Arkansas.

The prototype device, called ACONE 1.0, has built-in software that activates its two cameras only when a likely bird flies into view. It can rule out reflections and clouds and can recognise certain known traits of the woodpecker, including its average flying speed.

right: ACONE 1.0 installed above water
[photo courtesy of Ken Goldberg and Dez Song]

Wireless, plugless

The invention of a power system that transfers energy across a room without a wire in sight could transform the house of the future. Instead of household appliances needing to be plugged into a socket via a cord, they could simply be switched on wherever they are in the room. Laptops, mobile phones, MP3 players and other portable equipment would recharge automatically just by being in the room—whether placed on a table or still in a hand-bag or jacket pocket. The system could even be used to keep robots going.

The prospect of wireless energy supply has been made possible by researchers at the Massachusetts Institute of Technology in the US. They have developed a system, called WiTricity, which sends power through the air by employing magnetic fields and resonance. An experimental prototype successfully lit up a 60-watt bulb from a distance of seven feet (2.1 m) without any physical link between the transmitter and receiver.

The system takes advantage of resonance, the rate at which objects vibrate when energised. It has long been known that objects with the same resonant frequency can exchange energy—what would be happening if an opera singer hit a note that caused glass to shatter. By making the transmitter and receiver resonate through magnetic fields, at the same frequency, the researchers, led by Professor Marin Soljacic who was behind the innovation, created a means to transfer power.

By using non-radiating magnetic fields, the researchers hit upon a system that would only transfer power between the transmitter and receiver. Electromagnetic radiation, of which radio waves are a type, sprays energy in every direction. This works for transmitting data, but is too wasteful for a wireless electrical supply because virtually none would reach the intended target. Toothbrush chargers and a handful of other products make use of electromagnetic fields to transfer energy, but they only work effectively at close range.

By contrast, the WiTricity system works across several feet, and direction-finding technology is unnecessary—it does not matter if furniture gets in the way because the transferred power will pass right through. Health risks are thought to be unlikely, since the body is understood to be unaffected by magnetic fields. And because the system could reduce the number of batteries used in portable equipment, it is hoped it will slash the quantity of toxic pollutants that reach the environment from batteries each year.

The experimental unit was formed of two copper coils, each two feet (60 cm) in diameter. One was attached to the power source, the other to a light bulb. Having proved the system works, the MIT team expect to refine it to improve the 40 per cent efficiency with which power is transferred, and hope to develop it into a commercial product by 2012. Chief among the needed improvements are coils a more practical size for the home and extension of the distance with which power can be transmitted efficiently.

Objects with the same resonant frequency can exchange energy—what would be happening if an opera singer hit a note that caused glass to shatter

right: 60-watt light bulb being lit from seven feet (2 m) away, with the lower image demonstrating successful lighting with an obstruction [photos courtesy of Aristeidis Karalis, Marin Soljacic and AAAS]

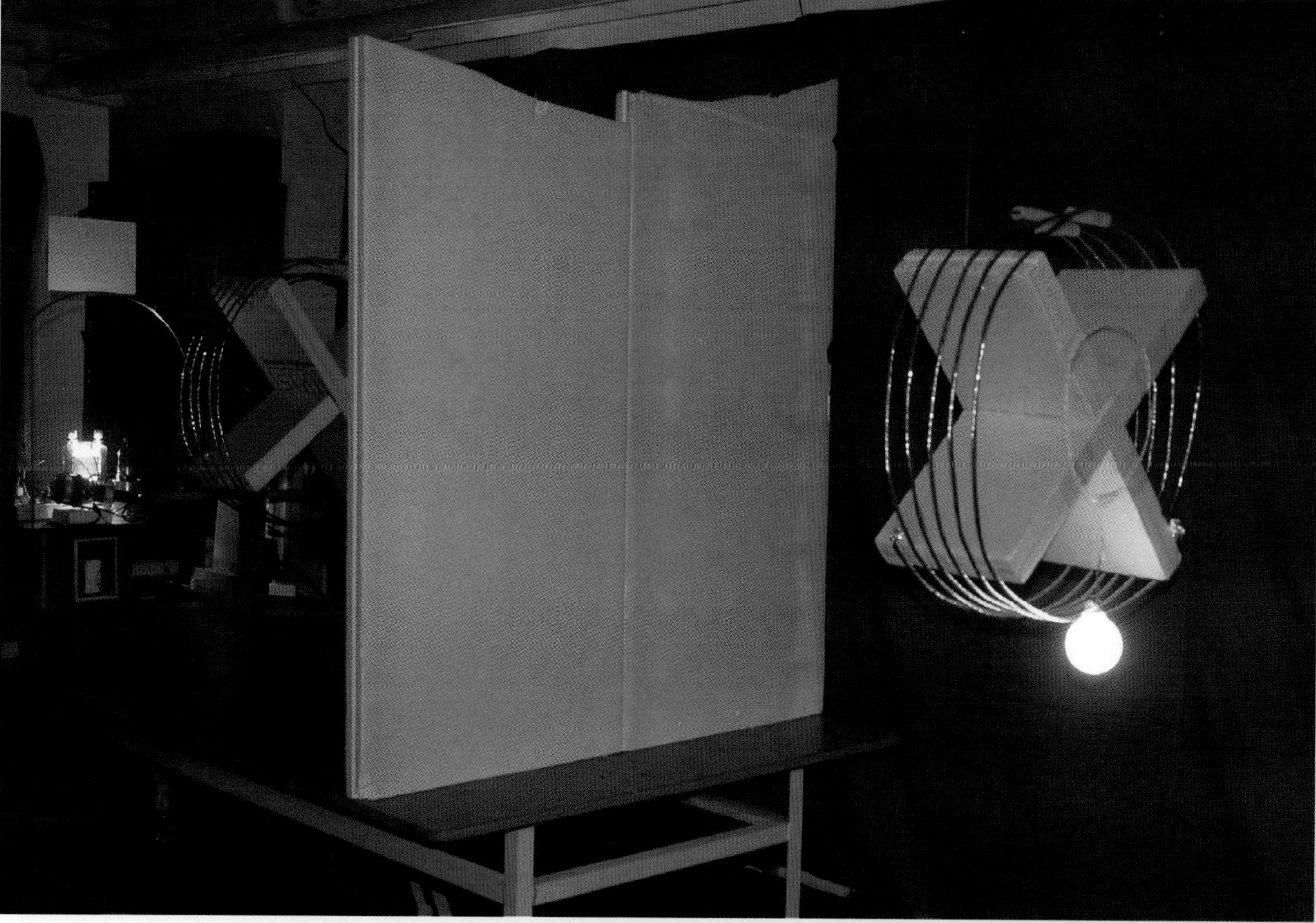

Micro power

The prototype generator draws electric current from ultrasonic waves, blood flow and mechanical vibration

A tiny generator has been developed, designed to provide power to nanodevices implanted into the human body that monitor health or carry out treatment. It is designed to be a self-sustaining source of electricity, with the prototype drawing electric current from sources including ultrasonic waves, blood flow and mechanical vibration.

The generator, devised by researchers led by Professor Zhong Lin Wang at the Georgia Institute of Technology in the US, consists of hundreds of zinc oxide nanowires. Each nanowire is spaced half a micron apart—one two-thousandth of a millimetre—and when bent by vibrations or waves, produces small electrical charges drawn off by silicon electrodes to supply direct current. The supply of four watts per cubic centimetre should be well within the requirements of nanodevices, including nanoscale robots, as they are developed for insertion into the body where batteries or refuelling would be impractical.

left and below: Zinc oxide nanostructures synthesised by a vapour-solid process for applications in nanogenerators and nanosensors [Images courtesy of Zhong Lin Wang, Georgia Institute of Technology]

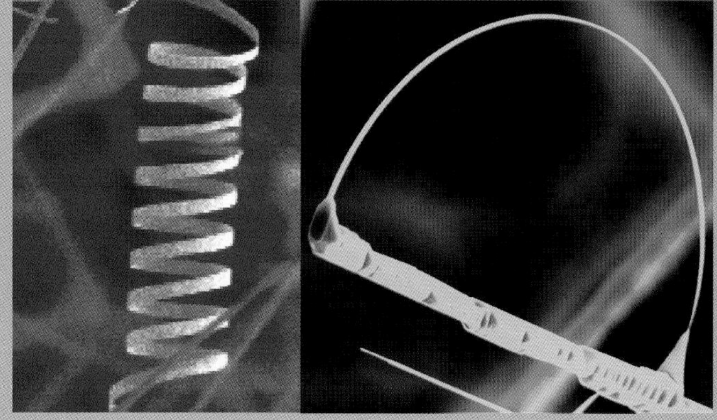

Deceiving

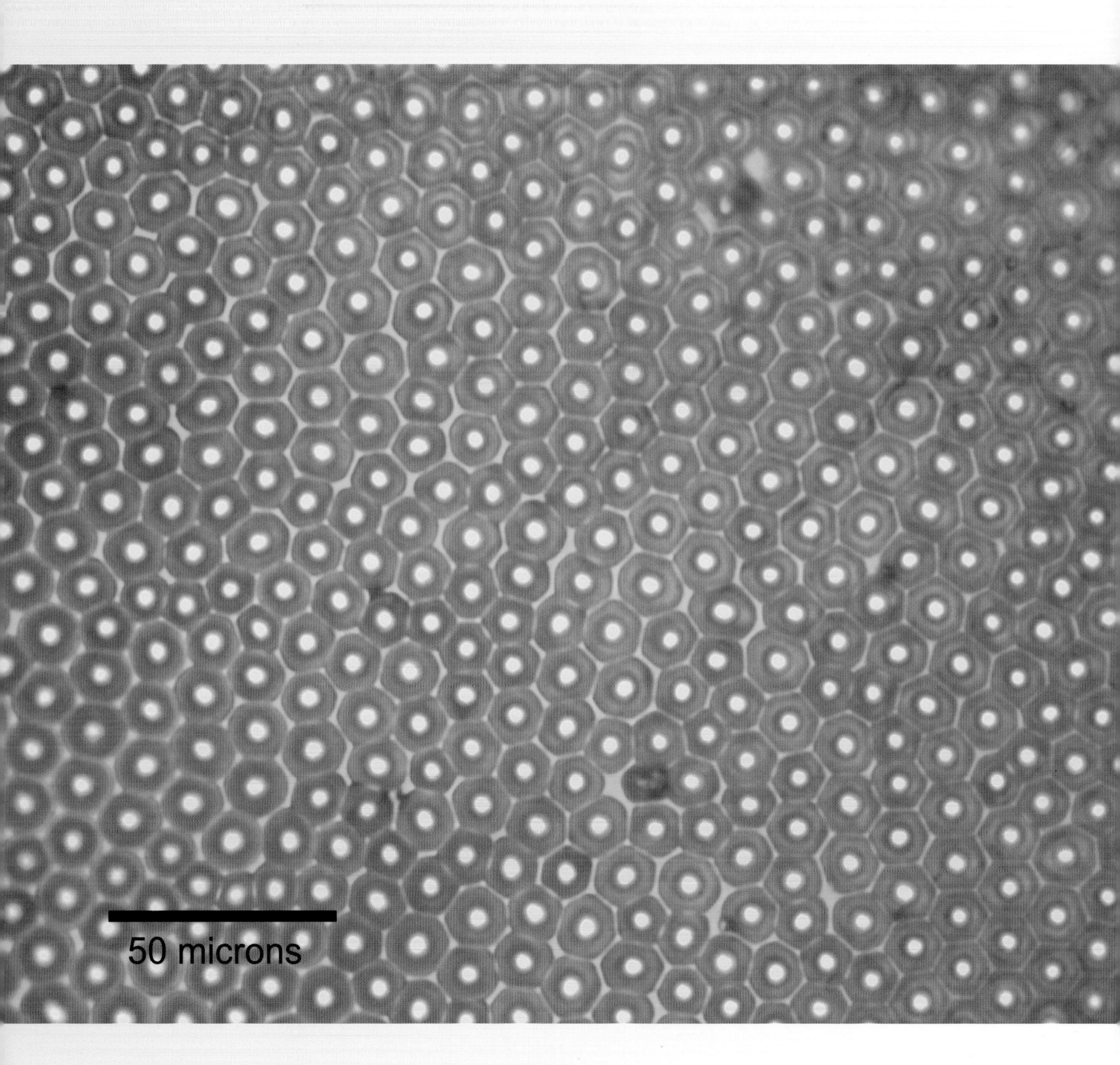

the eye

The art of the perfect mobile phone picture may lie in the dimples of a beetle's wing casing. New World scarab beetles have been found to reflect a "purer" light than most objects, and the lessons learned from the construction of its shell are expected to lead to better quality laptop and mobile phone screens.

The iridescent Costa Rican beetles, *Plusiotis boucardi*, deceive the human eye into seeing them as green. Observations with microscopes have shown that they are instead a honeycomb pattern reflecting red, yellow and green, and the honeycomb pattern with indentations a fraction of a hair's thickness is thought to be the key to the deception. The shells have no actual pigment and fibres in the shell are arranged so that they channel light waves through the outer layer, except for precise shades of red, yellow and green. While most colours of the light are absorbed, the blocked red, green and yellow shades are reflected back in unusually pure forms.

Researchers studying the beetle hope to harness the structural qualities of the shell casings, formed of fibres, to improve the performance of liquid crystal displays (LCDs). Dr Sharon Jewell, who led the research at the University of Exeter in the UK, believes that if LCDs can mimic the beetle shells, colours will be sharper and the energy needed to produce them will be reduced. In particular, it could do away with the need for LCD screens to incorporate light filters.

She said the crucial factor in the beetle shell is thought to be a dimple at the heart of each of the hexagonal structures that form the honeycomb pattern. The dimples are so small that they are only a tenth of the thickness of a human hair. Once scientists have worked out how exactly the wing casings channel light waves and separate them out, they expect to identify new techniques to improve LCD screens in laptop computers, mobile phones and even televisions.

left: Image detailing the *Plusiotis boucardi* beetle's honeycomb pattern on the central region of its back, magnified 50 times
right: *Plusiotis boucardi*
[images courtesy of Dr S. A. Jewell, University of Exeter]

The iridescent beetles deceive the human eye into seeing them as green

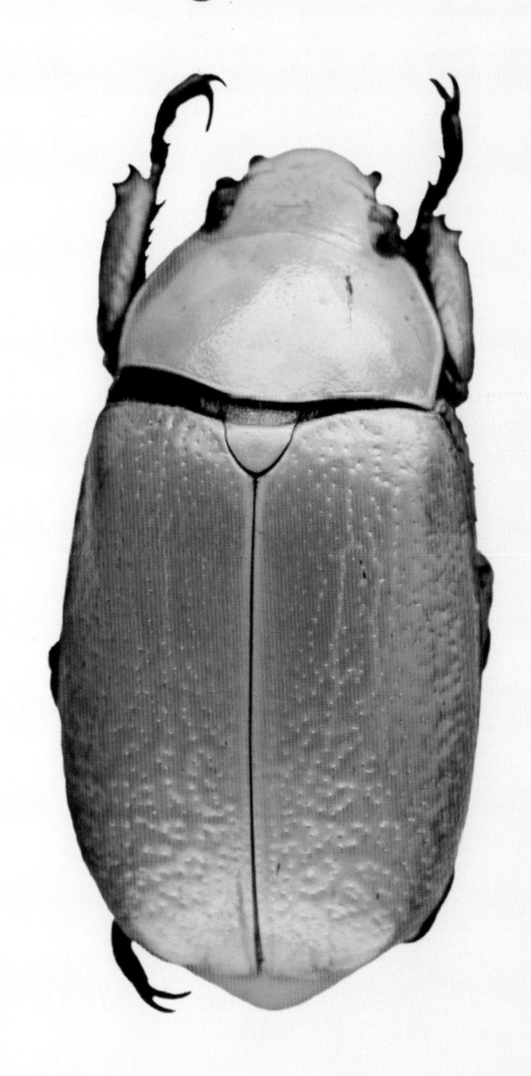

Fifth-column mosquito

A recently developed genetically modified mosquito holds the promise of eradicating or dramatically reducing malaria. A cure or prevention for malaria is one of the holy grails for science, with the disease killing up to 2.7 million people each year and infecting as many as 500 million. Genetic modification is increasingly seen as a potential solution to the malaria problem, for which the financial cost is counted in billions rather than millions of pounds or dollars annually.

The first GM mosquitoes were developed in 2000, but were out-competed by other species of the fly and would have died out in the wild rather than stop the spread of the disease. A GM, or transgenic, mosquito created by researchers in the US and Brazil, however, has been proven in laboratory conditions to be able to out-compete naturally occurring species infected with the malaria parasite. By altering the genetic makeup of the mosquito, researchers were able to make it resistant to malaria by the production of a protein, called SM1.

In the laboratory the transgenic variety, which were also modified to have bright green or red eyes to make it easier to distinguish them from other versions, increased from 50 per cent of the mosquito population to 70 per cent after nine generations. In the wild it would still bite humans and other animals to suck their blood, but, if the introduction went as planned, it would not infect them with malaria. Moreover, if the laboratory tests were replicated, it would take over habitat from naturally occurring mosquitoes, which carry the infection and suffer a reduction in breeding capacity when they are infected.

The successful lab experiments led by Johns Hopkins Malaria Research Institute in the US are only a step towards a solution, albeit one with enormous possibilities. More research is needed before GM mosquitoes can be released into the wild in the hope of reducing the impact of the disease.

Among the drawbacks is the finding that, although breeding capacity and lifespan of the transgenic mosquito were better than in previous experiments, it was only on an equal footing with non-GM mosquitoes that were free of the parasite. In the wild only a small proportion of mosquitoes have the malaria parasite, so the modified variety would be competing for the most part against noninfected flies that would be harder to displace, thus slowing the takeover.

Furthermore, the species of mosquito and parasite used in the experiment were not those that cause the most damage to the human population. The parasite chosen was *Plasmodium berghei*, which is an ideal variety in laboratory conditions but does not infect humans. The *P. falciparum*, however, is the most dangerous of the strains that do infect people. Access to the *P. berghei* parasite for the tests was through feeding on infected mice.

Similarly, the mosquito used for the experiments was *Anopheles stephensi*, which does pass malaria to humans, but is not the most dangerous to humans. That accolade goes to *A. gambiae*, which is the main source of infection in Africa.

A further obstacle is the concern that would almost certainly be raised by environmentalists worried by the level of human interference with the natural cycle of the environment. Biological solutions to problems have been tried in the past but, as the cane toad in Australia has shown, such initiatives often create more problems than solutions.

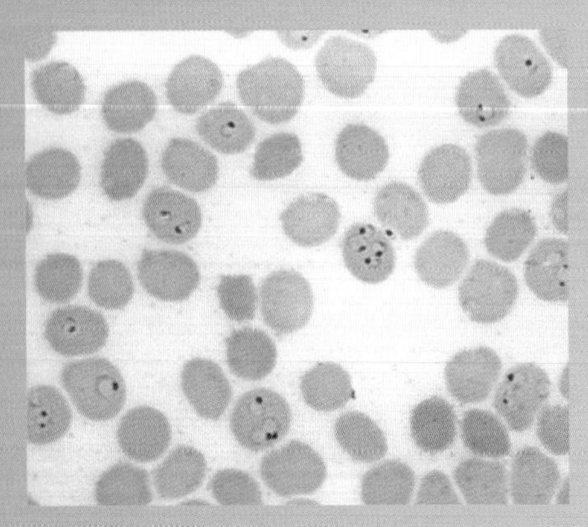

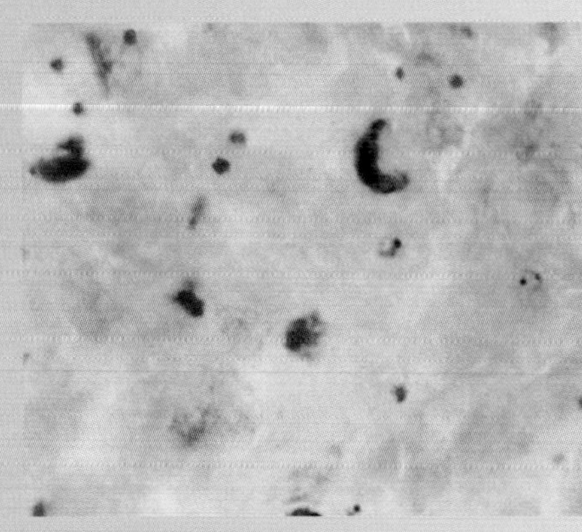

left: *Plasmodium falciparum* as seen through a microscope [images courtesy of Centers for Disease Control / Steven Glenn, Laboratory & Consultation Division]

Genetic modification is increasingly seen as a potential solution to the malaria problem, which kills up to 2.7 million and infects as many as 500 million people annually

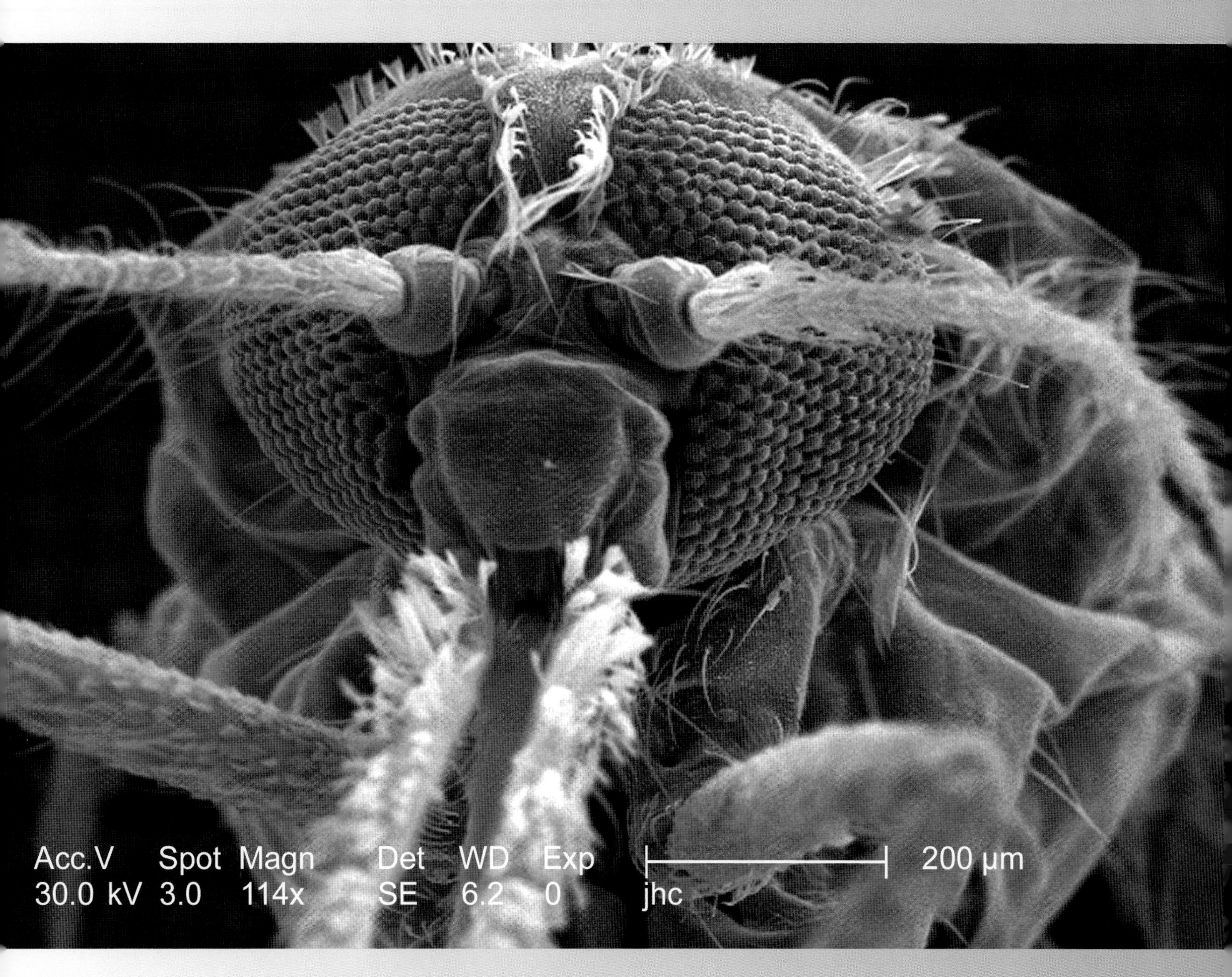

above: *Anopheles gambiae* mosquito close-up [image courtesy of Centers for Disease Control / Paul Howell]

Bionic eye

Millions of blind people could have their sight restored by a retinal implant. A device advanced enough to allow users to identify and distinguish between objects is being developed in the US and could be on the market by the end of the decade.

Researchers are confident enough in the technology that they predict that by 2015 the Argus device could have the capacity to allow blind people to see and recognise individual faces. It has been created to restore vision to people who have lost their sight through degenerative eye diseases that destroy photoreceptor cells in the retina.

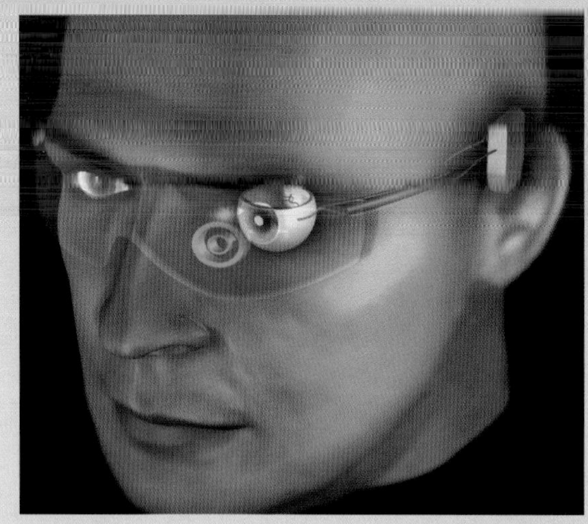

above: Depiction of the Argus device, which uses a small external camera to transmit images to an implanted retina chip [illustration: © 2005 Doheny Eye Institute]

The brain interprets the information sent to it by the device as shapes made up of light

Electrodes in a tiny platinum and silicon chip, fixed to the retina by a tack the width of a human hair, take the place of the photoreceptors and are used to stimulate ganglion cells, which in turn send information to the optic nerve. Information is passed to the chip from a radio receiver implanted close to the user's eye. A miniature camera within a pair of dark glasses records visually and transmits the data to the radio receiver. On a belt, the user wears a transmitter to process the information and a battery to power the equipment.

The first generation of the device was fitted with 16 electrodes and in tests proved more successful than the researchers had anticipated. Volunteers were expected to see just light and dark, but during equipment tests they were able to distinguish motion and the shapes of individual objects such as plates and cutlery.

Professor Mark Humayun of the Doheny Eye Institute, part of the University of Southern California, is one of the researchers involved in the US government-backed project and said an upgraded model with 60 electrodes, Argus II, is being developed for commercial use. Eventually it is hoped that a 1,000 electrode device, which should allow the user to recognise faces, can be built.

The brain interprets the information sent to it by the device as shapes made up of light. The shapes are built up piece by piece, just as a computer screen consists of pixels, and it may in time be possible to build up a colour picture of what is in the camera's field of vision. Tests with the 16-electrode chip were conducted with six volunteers who had lost their sight to the incurable eye condition retinitis pigmentosa, which affects one in 3,500 people.

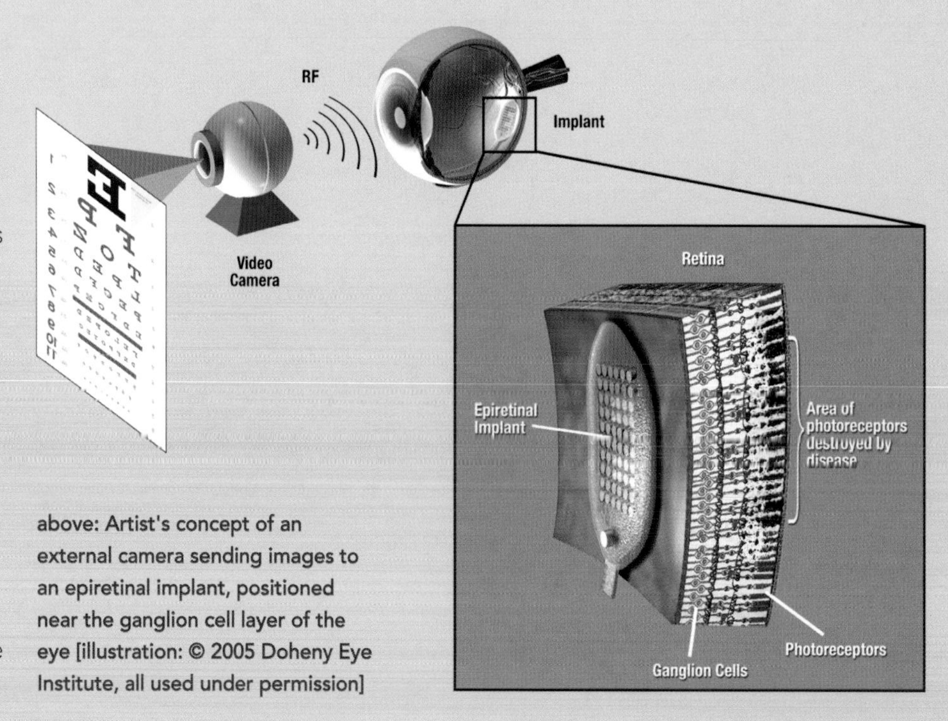

above: Artist's concept of an external camera sending images to an epiretinal implant, positioned near the ganglion cell layer of the eye [illustration: © 2005 Doheny Eye Institute, all used under permission]

Fat-burning pill

A pill that kick-starts the body into burning fat without the effort of exercising could be on the market within a decade, a leading researcher has claimed. Professor Ronald Evans of the Salk Institute in the US is confident that a once-a-day pill can be developed by 2013 to help tackle the growing problem of obesity.

The pill would work by tricking the body into thinking it is being given a rigorous workout. His team has succeeded in tests on mice in artificially activating PPAR-d, the molecular switch that controls the fat-burning process during exercise. By switching on the regulator, the mice became resistant to weight gain even when inactive and on a high-fat diet.

In 2004, Professor Evans led research that developed a genetically modified mouse, dubbed the Marathon Mouse, which was resistant to weight gain and had twice the physical endurance of other mice, being able to continue running for an hour longer.

The permanent genetic modification was carried out before the mice were born, an impractical solution to human obesity. A pill, however, could be used as a treatment for adults suffering human metabolic syndrome, which causes people to become obese, often leading to related health hazards such as heart disease, diabetes and high blood pressure.

In mice, a synthetic drug that mimics fat has been found to switch on the regulator without the need for genetically modifying the rodents first. Fat levels were reduced without affecting the endurance levels of the animals.

Professor Evans said that in a society where too few people exercise, sometimes because of existing medical conditions or because they are already too heavy to work out, an "exercise pill" would get rid of fat from the body while improving the quality of muscles and making the pill-popper healthier.

The pill would trick the body into thinking it is being given a rigorous workout

below: Three thermograms of an obese woman
[image: © Dr Ray Clark & Mervyn Goff / Science Photo Library]

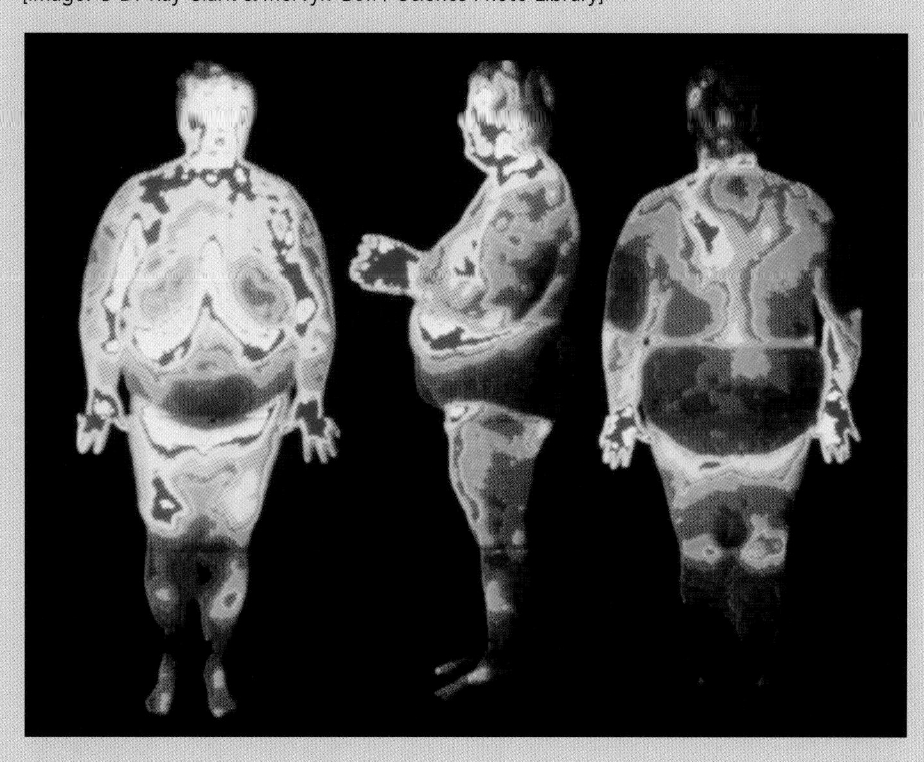

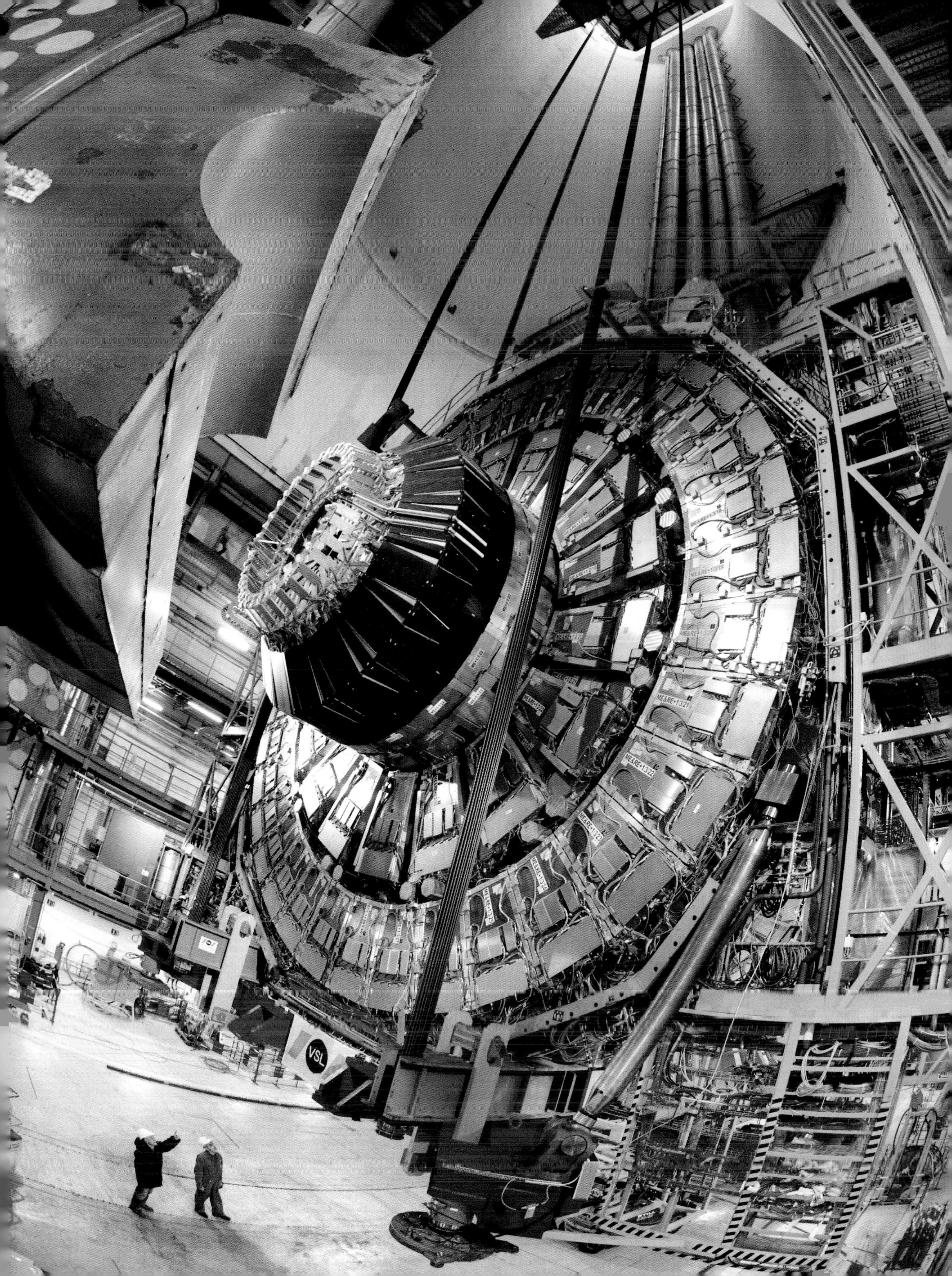
VSL

Unravelling the secrets of the universe

CERN, the world's largest particle physics laboratory located on the Franco-Swiss border, was designed to hunt down the smallest and most mysterious objects in the universe. Its Large Hadron Collider is suitably designed to be the biggest and most powerful particle detector in the world. Its main aim is to explain the rules that govern the universe, in part by identifying the tiny, unseen particles that exist in theory to explain its workings, but that have yet to be proven.

Among the first of its targets will be the Higgs bosun, a hypothetical particle that is thought to explain why matter has mass. It was proposed in the 1960s by Peter Higgs, now a University of Edinburgh emeritus professor, and is thought to cling to matter and drag on it as it winks in and out of existence.

The collider is a 17-mile (27 km) circular tunnel deep underground, in which two beams of either protons or lead ions, hadrons, are fired through tubes at close to the speed of light—0.999997828 times the speed of light to be more accurate. The beams cross at four points in the tunnel, smashing the particles into each other at a rate of 600 million collisions per second.

At these four intersections are arrays of recording and measuring equipment, which detect the energy and other particles given off by the collisions that replicate the conditions a billionth of a second after the Big Bang took place. The equipment at the intersections is built on a monumental scale. The biggest, Atlas, is more than 82 feet (25 m) tall and 150 feet long (46 m) and is designed to try to detect particles, including the Higgs bosun, and to find answers to what constitutes dark energy and dark matter, which form 96 per cent of the universe.

The Compact Muon Solenoid incorporates the largest magnet of its type, a superconducting Solenoid weighing 12,500 tons, with a field strength 100,000 times that of the Earth's. Among its aims is to detect previously unknown particles.

Each of the laboratory stations are expected to provide new information on what the universe is comprised of and how it fits together and why. The discoveries that are made will doubtless lead to further unknowns.

The collider is a 17-mile circular tunnel deep underground, in which two beams are fired at 0.999997828 times the speed of light

opposite: CMS experiment: lowering of the YE+1 endcap disc [photo courtesy of Maximilien Brice © CERN]
left: Aerial view of CERN [photo courtesy of AC Team © CERN]
overleaf: Central view of the ATLAS detector [photo courtesy of Maximilien Brice; © CERN]

LAYHER

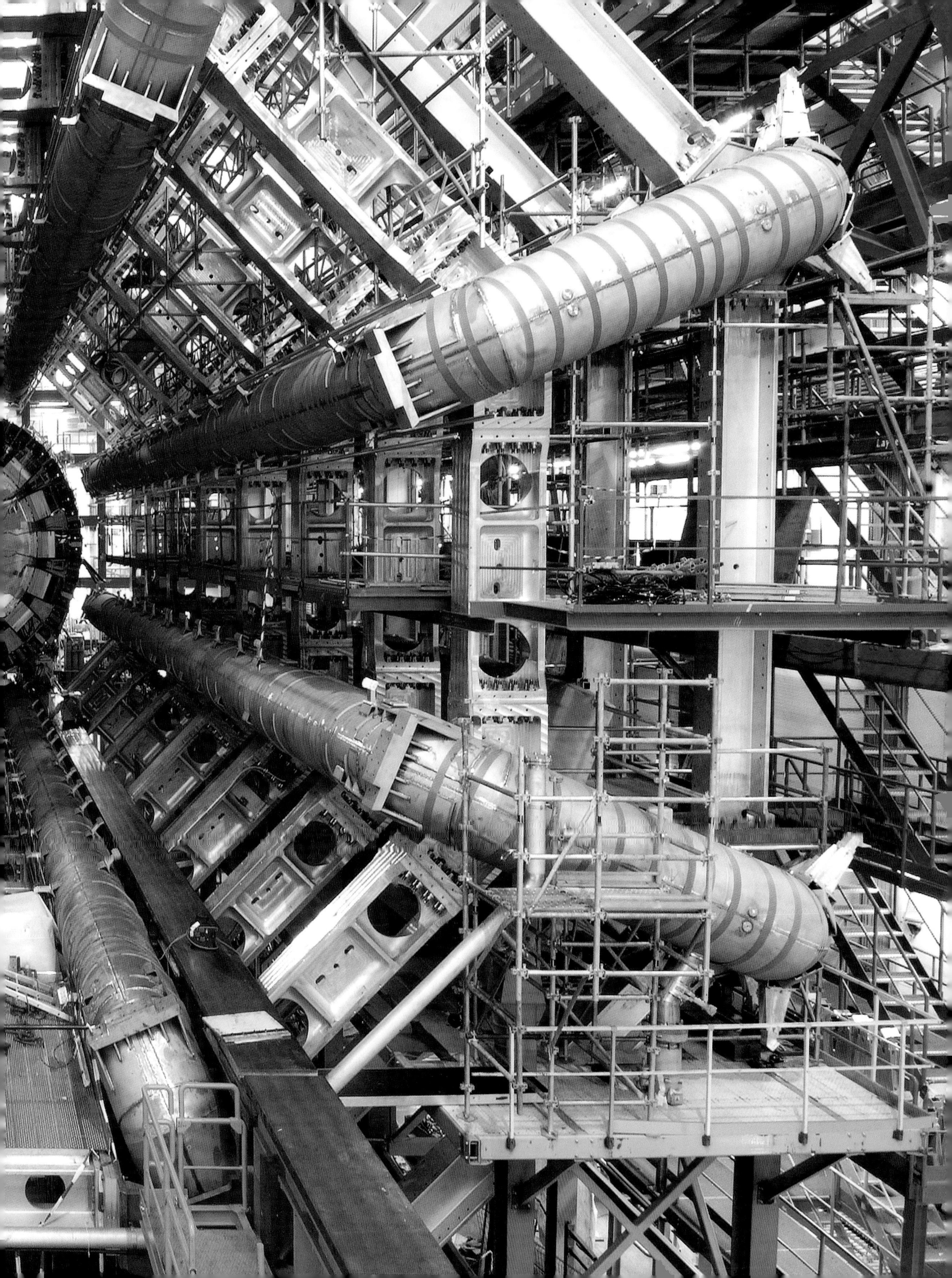

previous, clockwise from top left: Herring gull [photo courtesy of Grahame Madge], Antarctic ice fish and brittle stars overgrown by a yellow sponge [photo: © Julian Gutt / Alfred Wegener Institute for Polar and Marine Research], solitary sandpiper [photo: © Brian E. Small / VIREO], Golden-fronted bowerbird [photo courtesy of Bruce Beehler, Conservation International], *Bulbophyllum masdevalliaceum* [photo: © WWF / Wayne Harris], Bornean clouded leopard [photo: © WWF-Canon / Alain Compost]

New Species

Despite all the efforts of explorers and naturalists over the last few hundred years, new species of animals and plants are still being discovered. The vast majority are located in parts of the world, such as seas and tropical forests, where exploration has been hampered by inaccessibility.

Marine environments are so poorly charted by biological scientists that expeditions can identify previously unknown species by the dozen. Tropical forests, while better known, can offer similarly rich veins of discovery.

Yet even the most explored regions of the globe can yield new finds. In the UK—arguably the best understood environment in the world—new insects are identified every year or two, and plants occasionally turn up as well. The UK has even recently laid claim to a new species of bird, a result of scientific analysis that revealed a type of crossbill was, in fact, two types.

Mouse teeth

Discoveries of unknown mammals are extremely rare, but do happen, almost always in remote regions of the world such as South American rainforests. In 2006, however, an unknown species of mouse was found on the Mediterranean island of Cyprus and hailed as the first new mammal discovery in Europe for more than a century.

The Cypriot mouse, Mus cypriacus, was uncovered by French scientist Dr Thomas Cucchi of the University of Durham in the UK, who was studying teeth from Stone Age mice and comparing them with those of modern mice in attempt to establish when the house mouse arrived on the island. While searching for modern mice to provide specimens to examine, he realised he had caught a species that appeared different from all known varieties. Its head, teeth, ears and eyes were larger than those of mice known to live on the island, and genetic analysis confirmed that it was an entirely new species.

After comparing its teeth with those of fossil mice, Dr Cucchi concluded that Mus cypriacus had arrived on Cyprus and adapted to its environment thousands of years before people turned up. The discovery was all the more astonishing because Europe has been so well trawled by naturalists over the centuries, making it hard to believe that anything as big as a mouse remained undiscovered.

Animals such as the grey mouse would have been scampering through the undergrowth before mankind even reached Cyprus 10,000 years ago. This made the creature even more of a surprise, since all mammals endemic to islands in the Mediterranean—apart from two species of shrew—were driven into extinction by the advent of man and the animals that he brought with him, including the European house mouse.

Its head, teeth, ears and eyes were larger than those of mice known to live on the island

left: *Mus cypriacus* [photo courtesy of Anne Marie Orth; © CNRS]

Science on the menu

Exploring the few wildernesses that have so far escaped close attention from modern society is not just a matter of being prepared to go without the luxuries of hot, running water and baths for a bit. Researchers occasionally find their lives in danger when the wildlife they are trying to study turns on them.

Dr Enrico Bernard has experience in being quick on his feet to escape the attentions of meat-eating creatures—including a hungry cayman that snapped at his arm. Both a problem and joy of exploring regions where humans have rarely been seen before is, he said, that the local wildlife have little idea how to respond when people do turn up.

Animals in other parts of the world have been conditioned through bitter experience to, for the most part, regard humans as a danger best avoided. In largely unexplored areas such as the pristine, rainforest-covered state of Amapá in Brazil, however, few animals have ever seen a person and thus have not yet learned to fear people. Creatures such as tapirs, capybara and spider monkeys would, rather than flee, stop and stare when members of the Amazonia Project expedition, organised through Conservation International, came into view.

Such intimate contact with animals has its advantages, as expedition members could get closer than usual to the wildlife, with Dr Bernard describing it as "a scientist's heaven". But it also comes with disadvantages when the wildlife is big and hungry, and two of the project members had to spend a night sheltering in a hollow tree while a jaguar prowled about in hope of getting to them.

> In some of the areas being explored people are so rare that few of the animals have encountered or learned to fear them

Despite the occasional scare, the expedition returned from the region having recorded several animals that had never been seen before. Their tally, subject to review, included a tree rat, a bird, seven fish, two shrimps, eight amphibians and reptiles, and eight plants all previously unknown to science. The arboreal rodent discovered was the size of a guinea pig and of the *Makalata* genus. It was found to eat leaves and fruit.

The work carried out by the researchers was not limited to simply searching for unknown species. One creature they uncovered was an unusual four-fingered lizard, *Amapafaurus petrabactulus*, which had only ever been seen before in 1970. Researchers made comprehensive records of sightings of all the animals and plants that they encountered, and the data provides valuable information on the distribution and density of wildlife.

Amapá was previously considered to contain relatively few species for an Amazonian region, but observations by the expedition members belie the idea and suggest that it is in fact rich in variety. More than 1,700 species were recorded and more than 100 of them had never been seen in Amapá before. Moreover, those species that were known to be in the area were found in greater abundance than first suspected.

The leopard

that changed its spots

Perhaps the most spectacular recent discovery is that of the Bornean clouded leopard, the first new big cat identification in 200 years. Zoologists had known for more than a century that leopards lived in the forests of Borneo, but believed them to be the same species that lived in mainland Asia—mainly China, Nepal and Northeast India. However, analysis of the leopard's genes, combined with a detailed assessment of its fur pattern, revealed that, though more than a million years ago the Bornean variety had been the same as the mainland species, it had literally changed its spots in the intervening period.

About 1.4 million years ago, a land bridge that joined the islands of Borneo, Java and Sumatra to the Asian mainland disappeared under the sea, separating the clouded leopard populations. Over time, the clouded leopards on the islands—which had long been linked by land—evolved to take full advantage of their rainforest habitat, though they became extinct in Java in Neolithic times. Clouded leopards, which relative to body size have the longest canine teeth in the feline world, became known to Western science in 1821 when the British naturalist Edward Griffiths described them.

Dr Andrew Kitchener of the Department of Natural Sciences at National Museums Scotland has carried out a close examination of the leopards' fur and found clear differences between the mainland and island clouded leopards. On Borneo and Sumatra, the leopards—now named the Bornean clouded leopard, *Neofelis diardi*—have a double rather than single stripe along their backs and their fur is darker than the mainland species,

left: Bornean clouded leopard, *Neofelis diardi* [photos: © WWF-Canon / Alain Compost]

Neofelis nebulosa. Furthermore, on the Bornean cat the cloud patterns that give the mainly nocturnal predator its name are smaller than on the mainland species and contain more, better delineated spots.

The conclusion that the Bornean clouded leopard is a separate species, and not simply a subspecies as had been assumed for decades, was supported by genetic barcoding led by Dr Stephen O'Brien of the US National Cancer Institute. His team found more than three

They are as different genetically as lions are to tigers

dozen significant differences in the DNA—the genetic barcodes of the two leopards were so distinct that the declaration that they are separate species wasn't even a borderline decision. They are, in fact, as different genetically as lions are to tigers.

An estimated 5,000 to 11,000 leopards live in Borneo and a further 3,000 to 7,000 on Sumatra. The wide discrepancy between the top and bottom numbers reflects the difficulty in finding the creatures and how little research is carried out on them. Despite being the largest predator on Borneo, the reclusive and nocturnal leopards are seen so infrequently that the study of their fur patterns had to be based on 57 skins stored in museums and a handful of camera-trap photographs. Even so, researchers confessed that they were astonished no one had noticed what were to them obvious differences in the patterns on the cats' fur.

1.4 million years ago, a land bridge disappeared under the sea, separating the clouded leopard populations

left: Bornean clouded leopard, *Neofelis diardi* [photos: © WWF-Canon / Alain Compost]

Mystery moggie

Camera-traps usually help solve mysteries, but in the jungles stalked by the clouded leopard they have created one. A reddish cat-like creature, not much bigger than a domestic moggie, was photographed as it wandered along the forest floor at night in Borneo. Experts around the world have inspected the pictures, taken in 2005 by a camera-trap set up by researchers for the WWF, but no one has been able to identify the creature. Even local people had never seen its like before.

Conservationists are confident the animal is a new species. It is most likely a type of civet cat or marten, but it could be so different from anything else that it forms its own group. Analysis of the pictures suggests the creature is carnivorous, though some zoologists lean towards it being similar to a lemur, especially with its long and bushy tail.

More cameras have been placed to try to snap it again, and cage traps have been laid to try to catch a specimen of the animal—so far to no avail. It was pictured in the Kayan Mentarang National Park in Kalimantan, which is in the Indonesian section of the island, and conservationists fear it may become extinct before any more sightings are made.

No one has been able to identify the creature—even local people had never seen its like before

above: Reconstruction drawing (phantom drawing) of the possible new carnivore species [illustration courtesy of Wahyu Gumelar; © WWF-Indonesia]

Jungle punk

The world's first new primate for 20 years turned out to be even more surprising than its Mohican haircut after it was shown to be from an entirely new genus of monkeys. The kipunji was initially called the highland mangabey when it was discovered in a mountainous region of southern Tanzania in 2003 and 2004, and when first described in 2005 by two separate research teams it was given the scientific name *Lophocebus kipunji*.

But morphological and genetic analysis later revealed that it was more closely related to baboons than mangabeys, and required a new genus to be created to accommodate it in the classification system. The last time a new genus of primate was created was in 1923 when the Allen's swamp monkey, first discovered in 1907, was recognised as being a genus apart.

The genus attributed to the kipunji, after research led by Tim Davenport of the World Conservation Society, was derived from Mount Rungwe, where the first colonies of the creature were located. It is now officially known as *Rungwecebus kipunji*.

Having gone so long without realising the existence of the species, zoologists have been keen to learn as much as possible about the kipunji. Its most striking feature is a crest of long, upstanding hair, which has been likened to the Mohican hair style beloved of 1970s punks. With this are long whiskers and a thick coat that is thought to allow it to withstand the cold temperatures in its high-altitude habitat. Another unusual feature about the kipunji is its "honk-bark" call, a loud and low-pitched noise that is unlike the call of any other monkey.

Its most striking feature is a crest of long, upstanding hair, which has been likened to the Mohican hair style beloved of 1970s punks

At least 16 colonies, in groups of 30 to 36 adults, have been identified in the Rungwe-Livingstone forest and the Ndundulu Forest Reserve. They live in trees on mountains at elevations of up to 7,874 feet (2,400 m).

While the kipunjis were unknown to science until a short while ago, they were well known by people living in the area, who hunt them both for food and as a form of pest control. Locals blame kipunjis for damaging crops, and the kipunji's body used to identify the new genus, after a detailed examination of its skeleton and DNA, was in fact found in a farmer's trap. Until then it had only been analysed through the use of photographs.

left: Kipunji monkeys
[photo courtesy of Tim Davenport / WCS]

New birds for old

Discovering new species need not always require scientists to double as intrepid explorers, hacking their way through the undergrowth in remote and unfriendly terrain. Researchers working on a project designed to provide a genetic barcode for every animal in the world managed to find 15 new species among the birds of North America, some of the best-observed animal populations in the world.

Professor Paul Hebert of Guelph University in Ontario led the project, which aimed to analyse the DNA of 643 species of birds in North America, about 93 per cent of the continent's birdlife. Once all 10,000 known bird species around the globe have been DNA barcoded in the same way, he suggested, a further 1,000 new birds are likely to be identified on the basis of what was found in this study. Equally, a number of supposedly known birds may be lost as separate species, as during the project researchers found that several species had been double or triple-counted. It turned out that one type of gull, for example, had been categorised as eight different species.

The research was the first full-scale attempt to draw genetic pictures of a large quantity of species, and shows that the techniques should work for other types of animals in other parts of the world. The DNA barcoding analysis demonstrated some of the weaknesses of more traditional taxonomy, in which animals are described on the basis of colour and internal and external shape. The gulls were particularly illustrative of the pitfalls of less technologically rigorous methods—especially those described in the days when discovery was a matter of risking life and limb, with naturalists making their finds thousands of miles from either civilisation or the identical birds' previously known territory.

Over the centuries, eight gulls had been described in different parts of the world and identified as separate species. Yet when they were assessed, their genes were found to be virtually the same. The California gull (*Larus californicus*), herring gull (*L. argentatus),* Thayer's gull (*L. thayeri*), Iceland gull (*L. glaucoides*), lesser black-backed gull (*L. fuscus*), western gull (*L. occidentalis*), glaucous-winged gull (*L. glaucescens*), and the glaucous gull (*L. hyperboreus*), should all be classified as the same bird or at the very least only as subspecies, according to their DNA.

Similarly, 28 so-called species were found to be genetic twins, meaning they should actually be classed as 14 species, and six were genetic triplets that should be regarded as only two species. The snow goose and Ross's goose were examples of twin birds, the research revealing they are 99.8 per cent genetically identical.

Those birds that were discovered to be new by the researchers were almost visually indistinguishable from other species, but clearly different genetically. Solitary sandpipers were one of those that were in reality two types of bird; researchers suggested a second species should be recognised with the name cinnamon sandpiper. The two birds, the barcoding revealed, had been separate types for 2.5 million years.

The barcoding research was carried out as part of the Barcode of Life project, which hopes to collate 10 million records of 500,000 animal species by 2014.

The DNA barcoding analysis revealed some of the weaknesses of more traditional taxonomy

opposite: Solitary sandpiper [photo © Brian E. Small / VIREO]

Over the centuries, eight gulls had been identified as separate species. Yet when they were assessed, their genes were found to be virtually the same

overleaf: Herring gulls, *Larus argentatus* fighting over fish in the sea
[photo: Graham Eaton / RSPB Images]
below: Herring gull
[photo courtesy of Grahame Madge]

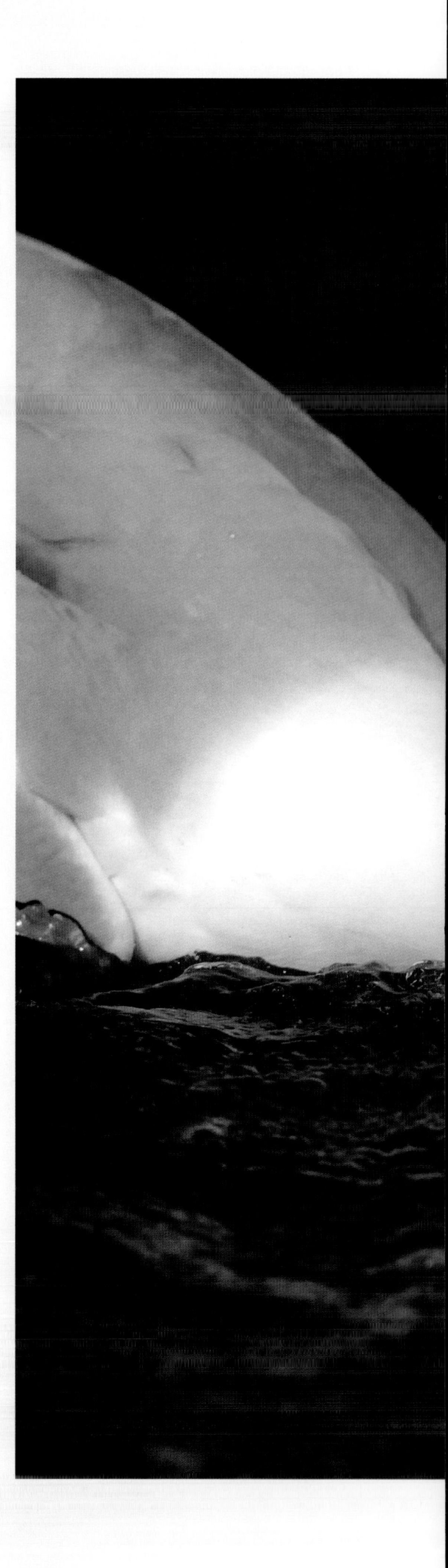

Long-distance licking

If a domestic cat were to have a tongue of the same proportion, it would be able to lap milk from a bowl placed two feet from its mouth

Simultaneously with the bird barcoding, the genes of 87 types of bats found in Guyana were being analysed, yielding the discovery that there were six more species than previously realised. Guyana's bats have been subjected to considerable study already, described in great taxonomical detail, yet six went unnoticed until the barcoding. If this trend is repeated when the more than 1,000 known bat species worldwide are tested, a further 50 species are likely to be uncovered.

But some bats are still being discovered the old-fashioned way—by traipsing out into the jungle to find them. This was indeed how the longest tongue proportionate to body size boasted by any mammal, and second only to the chameleon in the vertebrate world, was netted and brought back to laboratories for further study.

Unlike other bats, which have tongues starting at the back of the mouth, the tube-lipped nectar bat, *Anoura fistulata*, has a tongue so long that it has to pass back along the neck and into the thoracic cavity, where it rests between the heart and the sternum. Tube-lipped nectar bats are only two inches long, yet their tongues stretch out 50 per cent longer than their body length, with one of the bats even boasting a reach of 3.4 inches (8.6 cm) with its tongue. By contrast, if a domestic cat were to have a tongue of the same mammoth proportion, it would be able to lap milk from a bowl placed two feet (0.6 m) from its mouth.

The reach was assessed in experiments led by Nathan Muchhala of the University of Miami in the US, in which the tube-lipped and two other nectar-drinking bats were trained to stretch for sugar water with their tongues along see-through tubes. The outsized tongue developed to allow the bat, which lives in the cloud forests of Ecuador, to drink the nectar of the flower of the *Centropogon nigricans* plant.

Several nectar-loving bats live in the region in the Andes, but only the tube-lipped nectar bat can reach far enough with its tongue to reach the flower's nectar. It is believed that the trumpet-like, pale green flower and the bat coevolved, and *Centropogon nigricans* is the only plant known to rely solely on a bat for pollination.

Such a long-tongued bat might have been expected to grow a long snout to help house it when not in use. In fact, the bat's snout is comparatively short, making it easier for the creature to snap up insects as a protein supplement to its diet.

opposite: *Anoura fistulata*
below: Illustration comparing the tongue length of *Anoura fistulata* and other nectar bats [photo and illustration courtesy of Nathan Muchhala]

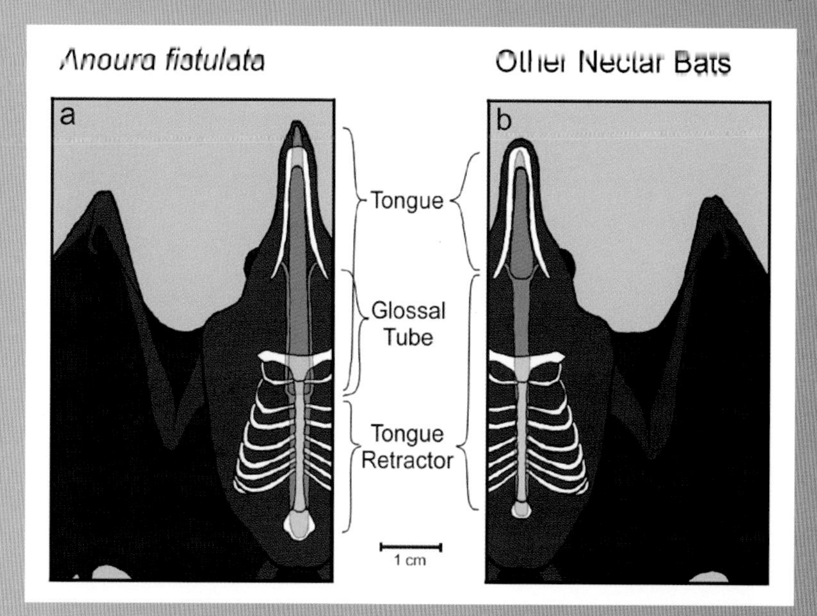

Treasure Island

Borneo's rainforests, as attested by the Bornean clouded leopard, are another of the world's wildernesses where exploration is still providing science with a string of new discoveries. The area's description by Charles Darwin as a "great wild untidy luxuriant hothouse" remains true today and, combined with its scale, enables both big and small species to stay out of sight. Efforts to identify the animals and plants within the forests, especially those that are under threat of being cut down, have been stepped up in recent years amid concerns about long-term conservation.

The governments of Indonesia, Brunei Darussalam and Malaysia have recently made several commitments to protect the rainforests from loggers, especially in the region known as the Heart of Borneo, but conservationists still fear trees are being cleared too quickly. The island boasts incredible diversity, including the highest number of plant species in the world, and plenty of new animals and plants are still turning up as botanists and zoologists try to map the area's biodiversity.

From 1994 to 2004, according to figures collated by the World Wildlife Fund, new species were being discovered so fast that three were being identified every month. In the ten years to 2004, at least 361 new species were discovered and since then more than 50 others have been identified, including a snake that can change colour. The snake, later called the Kapuas mud snake and classified within the *Enhydris* genus, was a reddish-brown when it was picked up by Dr Mark Auliya, of Zoologisches Forschungsmuseum Alexander Koenig in Germany, and placed in a bucket. When he looked at the snake again a few minutes

Changing colour is an ability boasted by only a handful of reptiles, most famously the chameleon, and is extremely rare in snakes

later, its colour had changed to white. Changing colour is an ability boasted by only a handful of reptiles, most famously the chameleon, and is extremely rare in snakes. Two specimens were found, each about a foot and a half (0.5 m) long, close to the Kapuas River in the Betung Kerihun National Park.

Six Siamese fighting fish, a catfish with goofy teeth and a sticky belly that allows it to stick to rocks in fast-moving streams, and a frog with an extra wide head were among other animals pinpointed by researchers.

Of the new plants, which included 16 types of ginger, one of the more surprising discoveries was a species with white flowers but just a single leaf. While new to science, the plant, *Schumannianthus monophyllus*, had long been known to the native Iban peoples, who use the leaf to wrap sticky rice at festivals.

With an estimated 15,000 plant species, 222 mammal, 420 bird, 100 amphibian, more than 400 fish and at least 150 reptile, and a landscape that is widely described as "inaccessible", it is perhaps reasonable to expect many hundreds or even thousands of plants and animals are still awaiting discovery.

left: Kapuas mud snake
[photo courtesy of Mark Auliya]

Tiny vertebrates

Among the 30 new species of fish recently discovered in Borneo is a species so small that it is the second-smallest vertebrate on Earth. *Paedocypris micromegethes* is just over a third of an inch (8.8 mm) long and lives in shady areas within the streams and pools of acidic peat bogs. The smallest known vertebrate is its cousin, *Paedocypris progenetica*, from the neighbouring island of Sumatra. Its discovery was announced in early 2006.

Paedocypris progenetica measures just less than a third of an inch (7.9 mm) long and four one-hundredths of an inch (1 mm) wide. Despite being a member of the carp family, it is so small and see-through that it is easily mistaken for aquatic larvae. Its habitat is murky water of the forest peat swamps that, with a level of acidity measured at three on the pH scale, is at least 100 times more acidic than rainwater. The swamps were thought to be too acidic to harbour much life, but recent research has shown them to be a surprisingly popular habitat, albeit not always for species already familiar to zoologists.

As well as being extraordinarily small, the fish, found by Dr Maurice Kottelat and Dr Tan Heok Hui from the Raffles Museum of Biodiversity Research in Singapore, has caught the attention of scientists due to odd features found on its body. The male, which is slightly bigger than the female, has a pelvic fin on the underside of its body and, just in front of it, a collection of muscles that appear to be able to grip. It seems most likely that the fin and the muscles are designed to grip the female during mating. Equally puzzling was the realisation that most of the top section of the skull is missing, leaving the brain exposed. Dr Ralf Britz of the Natural History Museum in the UK analysed the structure of the fish and concluded it was "one of the strangest fish" he had ever seen.

The specimen's peculiar fin and exposed brain led Dr Britz to conclude it was "one of the strangest fish" he had ever seen

Orchid hotspot

Hitherto unexplored rainforests have proved fertile hunting grounds for botanists in Papua New Guinea, where there is a profusion of orchids. More than 3,000 orchid species are known on the island and a series of expeditions to the Kikori region have uncovered even more. Botanists visiting the area as part of exploration overseen by the WWF collected 300 species from 1998 to 2006, of which eight were found to be new to science and a further 20 were suspected of being new, subject to confirmatory research.

More than 3,000 orchid species are known to thrive in Papua New Guinea

clockwise from left: The rare orchid species *Bulbophyllum masdevalliaceum*; the newly discovered *Taeniophyllum*; the recently discovered *Cadetia*; the newfound *Cadetia kutubu*; and the rare *Trichoglottis sororia* [photos courtesy of and © WWF / Wayne Harris]

Ecosystem creation

The destruction of an ice shelf in the Antarctic has given marine scientists a unique opportunity to watch the birth of an ecosystem. The breakup of the Larsen ice shelf, which took place in two stages in 1995 and 2002, has opened up the sea floor to a host of creatures that could not survive there when the sea was frozen above.

A survey of the seafloor as part of the wider Census of Marine Life revealed the first signs that different animals are moving into the area, which was once covered by ice up to 650 feet (198 m) thick. Sea squirts are among the most notable new tenants of the seabed. In places, they have become the most numerous organisms and their presence is a sign that the old occupants are being displaced now that conditions have changed.

The disappearance of the 3,861-square-mile (10,000 sq km) Larsen ice shelf led to the return of phytoplankton and zooplankton, providing the basis of the food chain. With them have come larger animals such as krill, seals and whales. The increased quantity of life near the surface means more nutrients sink down to the seafloor, giving a different range of creatures a chance to colonise it.

Growth of life in such cold conditions is slow, but over the next few decades the seafloor, which is up to 2,800 feet (853 m) below the surface in parts, is expected to become much richer. During the survey of the area and open waters in the northwest region of the Weddell Sea, an estimated 1,000 species of animals and plants were recorded, 20 of which were new to science.

Among the new species were 15 amphipods. One was a giant at four inches (10 cm) long, which may not seem much by human standards, but for shrimp-like creatures it is huge and one of the biggest yet seen. Also found were four cnidarians, a lifeform related to jellyfish and corals, and a sea anemone in a symbiotic relationship with a snail—it provided the protection, the snail the locomotion.

Animals are moving into an area that was once covered by ice up to 650 feet thick

below: Sea cucumbers from the Larsen B area
opposite, top row from left: Seafloor near Seymour and Paulet Island; sea star from Larsen A
opposite, centre row from left: Sea Anemone from Larsen B; corals from Larsen B; sea squirts from Larsen A
opposite, bottom row from left: Seafloor near Seymour and Paulet Island; Antarctic ice fish and brittle stars overgrown by a yellow sponge [all photos © Julian Gutt / Alfred Wegener Institute for Polar & Marine Research]

Living fossil

above: A Jurassic shrimp, *Neoglyphea neocaledonica*, found in the Coral Sea [photo courtesy of B. Richer de Forges, © 2006]

A crustacean thought to have died out 50 million years ago was dubbed the Jurassic shrimp after being found alive and well in the Coral Sea.

Marine biologists from France caught the shrimp, *Neoglyphea neocaledonica*, as they explored an underwater plateau in the waters off Northeast Australia. It looked like a cross between a shrimp and a lobster and, like the coelacanth discovered in 1938, was a species only previously seen in the fossil records.

Only one other crustacean of its type, *Neoglyphea inopinata*, has been discovered, but the newer animal has much larger eyes, suggesting it is a hunter dependent on good sight.

The shrimp, thought to have died out 50 million years ago, had only previously been seen in fossil records

Walking sharks

Sharks that walk on their fins were among 52 new species discovered in surveys of the so-called Coral Triangle, an area covering 70,000 square miles (181,300 sq km) of the Indian Ocean. Within this is the Bird's Head Seascape at the northwestern end of the Papua province of Indonesia, home to more than 1,200 species of fish and almost 600 species of reef-building coral—about three-quarters of the total number throughout the world. It is so rich in life that in places an area twice the size of a football pitch contains four times as many species of reef-building coral than the entire Caribbean Sea.

Two types of epaulette sharks that use muscular pectoral fins to walk along the seabed at night were among the newly discovered creatures. They can grow to four feet long and it is thought the ability to walk allows them to keep close to the seafloor where they hunt, hauling themselves between tight spaces in the reefs in search of small fish, crabs and snails. Because they are moderately small as sharks go, sticking close to the bottom helps them avoid bigger predators.

In among this rich variety, scientists from Conservation International discovered 20 new coral species, 24 new fish and eight previously unknown mantis shrimp during three surveys in 2001 and 2006. A species of "flasher" wrasse was a particularly interesting find. The males, which have brown as their default colour, flash brilliant pink or yellow to attract mates.

Their ability to walk allows them to keep close to the seafloor while hunting small fish, crabs and snails

below: Shark walking along the seafloor at Triton Bay, West New Guinea [photo courtesy of Gerry Allen]

The quantity of new types of sponge that they found was the largest for the North Atlantic since the days of the Victorian collector

Sponges

Almost as colourful, though for different reasons, were a variety of bright red and yellow sponges discovered off the coast of Ireland. Divers from the Ulster Museum discovered up to 47 new species of sponges about 100 feet beneath the waves near Rathlin Island, off County Antrim. The finds, 28 of which have been confirmed as new species as the rest await analysis, add a significant number of species to the 350 sponges already known to live in the waters of the British Isles.

Being so deep under the surface of the sea, the sponges, which are filter-feeding animals and among the most ancient of creatures, live in virtual darkness. In good sunlight the reds and yellows chosen by the sponges would be easily visible, but in the darkness of their natural habitat the colours are the hardest to see. Being as good as invisible—until divers with bright lights turn up—the sponges are perfectly camouflaged and any predator hoping for a sponge snack has to sniff them out.

The new sponges range in size from less than an inch (2.54 cm) in diameter to about three inches (7.62 cm). While they may sound diminutive, they are the bullies of the seabed, stealing all the best rocks from any anemones and bryozoans that happen to get there first. In shallower water they are out-competed by seaweed, and thus thrive in deeper water, where in some areas off Rathlin Island they cover up to 70 per cent of the rock surfaces.

Bernard Picton, the museum's curator of marine invertebrates, said they are the seafloor's dominant species in that part of the ocean. So rich is the area in sponges that during a six-week expedition in 2006, museum divers were able to locate 128 different species. The quantity of new types of sponge that they found was the largest for the North Atlantic since the days of the Victorian collector.

The new sponges are the bullies of the seabed, stealing all the best rocks from any anemones and bryozoans that happen to get there first

left: One of the 28 new sponge species discovered off Rathlin Island [photo: Ulster Museum]

Lost world

An ornithological riddle that had taxed researchers for more than a century was finally solved when conservationists set out to explore an area of the remote Foja Mountains of New Guinea. While Berlepsch's six-wired bird of paradise, *Parotia berlepschi*, was first described in the late nineteenth century, no one in the scientific community knew its home territory because all the bodies had been passed on by native hunters.

Several attempts were made to search for the bird, and all had failed until an expedition organised by Conservation International and the Indonesian Institute of Science set off for the pristine tropical jungle of the Foja Mountains. The team hit the jackpot on only the second day, barely having to get out of bed to find it—two of the birds wandered into the field camp and began courting, with a resplendent male performing a mating dance for a female. It was the first time a live male had been seen by Western scientists, let alone been seen during its mating ritual.

They had similar success with the golden-fronted bowerbird, which was known to be in the region and last spotted during an expedition in 1981, but had never been photographed during its mating display. This time it was pictured in all its glory beside the tower of twigs and other forest debris that it uses to impress the female.

While gathering specimens of 20 frogs and four butterflies that were hitherto unknown, the team discovered the first new bird species in New Guinea for 60 years, an orange-faced honeyeater, which has a distinctive patch of orange on its face. Equally exciting to the international team of researchers was a rhododendron flower, which at six inches (15 cm) across was thought to be the biggest in the world, and the golden-mantled tree kangaroo, *Dendrolagus pulcherrimus*, previously only found on a mountain in neighbouring Papua New Guinea.

To Bruce Beehler, one of the expedition leaders, the region was a lost world where many creatures, including the rare long-beaked echidna, were so unfamiliar with humans that the researchers were able to simply walk along and pick them up. It was, he said, about as close to the Garden of Eden as is likely to be found on Earth.

The creatures were so unfamiliar with humans that researchers were able to simply walk along and pick them up

opposite: Golden-fronted bowerbird
below: Bird of paradise [photos courtesy of Bruce Beehler, Conservation International]

Distinct accent

A dispute over the status of the Scottish crossbill rumbled on for more than 100 years—until researchers discovered it had a Scottish accent. The bird, *Loxia scotica*, was finally declared a species rather than a subspecies after analysis by scientists from the Royal Society for the Protection of Birds (RSPB) in the UK. Some ornithologists already regarded it as a separate species, but many, including bird scientists, remained doubtful until the bird's accent was assessed and found to be distinct.

The bird shares its forest habitat in Scotland in the UK with two other crossbills—the common crossbill, which has a small bill for extracting seeds from spruce cones, and the parrot crossbill, with a bigger beak for getting at seeds in pine cones. The size of its own bill lies in between the other two and allows it to gather seeds from several species of conifer.

Differences between the bills were insufficient to warrant declaration of the Scottish crossbill as a separate species, as were samples of DNA that showed close similarities between the three birds, all members of the finch family.

When the birds' calls were studied, however, it became clear that there were unmistakable differences and that the three crossbills would respond only to their own mating calls. The Scottish crossbill makes a "chup, chup" sound, whereas the common crossbill emits a "chip, chip" call and the parrot crossbill a deeper "kop, kop". Dr Ron Summers, who led the study for the RSPB, said the finding meant the UK finally had a species of bird found nowhere else in the world.

There were unmistakable differences in the birds' calls: the three crossbills would respond only to their own mating calls

opposite: Scottish crossbill male, *Loxia scotica*, in Scots Pine, Speyside, Scotland [photo: Danny Green / RSPB Images]

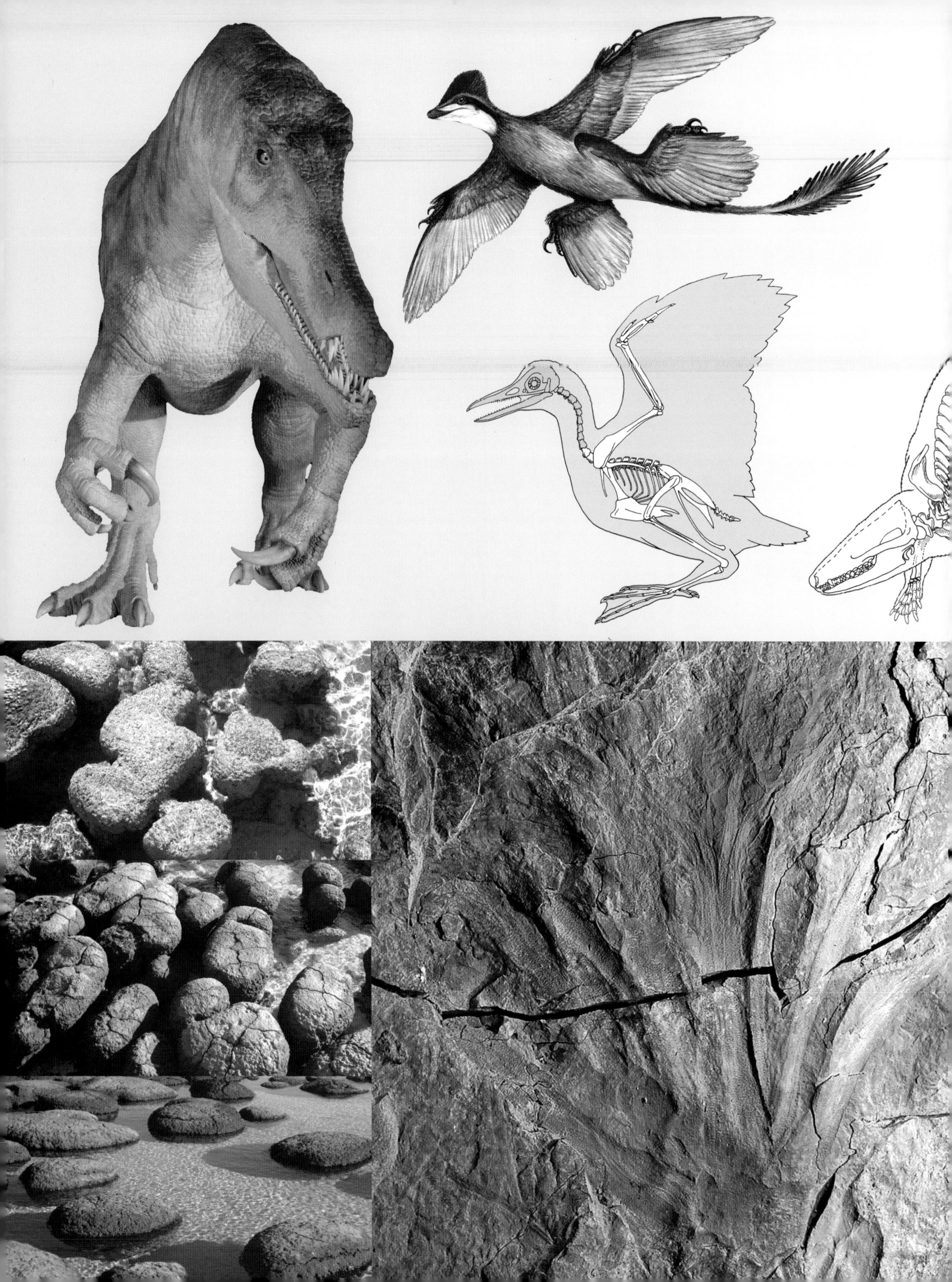

previous, clockwise from top left: A model of *Baryonyx* [photo: © Natural History Museum, London], illustration of *Microraptor gui* [© Andrey Atuchin / NHM], skeletal reconstruction of *Gansus yumenensis* [courtesy of Mark A. Klingler / CMNH], *Castorocauda lutrasimilis* skeleton reconstruction [courtesy of Quiang Ji, © Mark A. Klingler / CMNH], illustration of a *Oryctodromeus cubicularis* head [courtesy of Lee Hall, Montana State University], illustration of *Volaticotherium antiquus* [courtesy of Zhao Chuang and Xing Lida], a fossilised branch from the crown of a lycopsid tree [photo courtesy of Howard Falcon-Lang, University of Bristol], Wattieza tree fossil specimen [photo courtesy of William Stein, Binghamton University], three examples of stromatolites from Shark Bay, Australia [photos courtesy of Abby Allwood]

Life Gone By

Fossils provide a window to the past. Through them we can see what life looked like long before humans began to make records of the world around us. The view is, however, inevitably incomplete and paleontologists have often found themselves having to revise conclusions when a fresh fossil find is made.

In 1854 the Crystal Palace Dinosaurs, a series of models intended to give the Victorian public a glimpse of what the fearsome creatures looked like, opened to great acclaim and have remained a source of fascination to visitors to this day.

The model monsters in South London were a brave and innovative attempt at the first life-sized representation of dinosaurs, but it was soon realised that they were wildly inaccurate.

As more fossils were uncovered and the understanding of how they fit together improved, it became possible to build up a more accurate idea of what the animals looked like when they were alive. Similarly, fresh discoveries in Schoharie County, New York, in the United States, have given scientists a clear idea of what the first forests looked like.

The fossils dated to 385 million years ago, a time some 140 million years before the dinosaurs, when even amphibians hadn't managed to clamber out of water

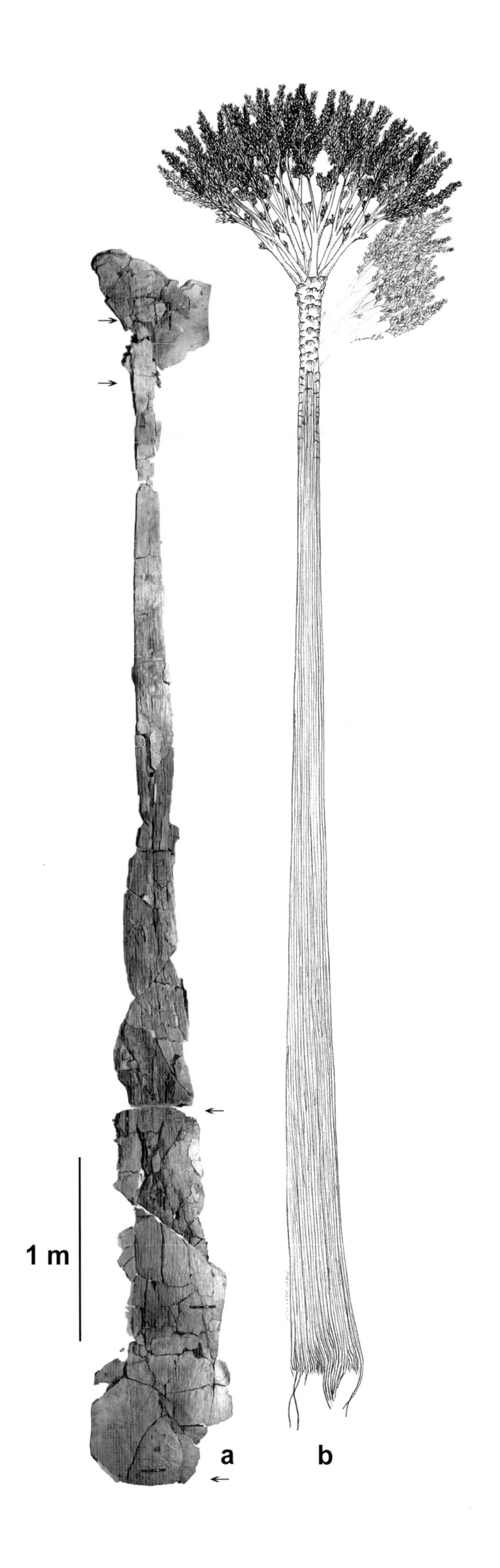

right: Wattieza fossil specimen and reconstruction
opposite: Wattieza fossil crown (a), reconstruction (b), fossilised branch (c) [images courtesy of William Stein, Binghamton University]

Humungous trees

Rock blasting at a quarry in 1870 uncovered the fossilised stumps of trees in Gilboa, New York, in the US. More than three dozen further stumps were discovered in the 1920s when the Schoharie Reservoir was being constructed, and subsequent analysis revealed that they formed the earliest known forest on earth. But with just the stumps preserved, the biggest being just over three feet (1 m), it was impossible to say what the leaves and branches looked like until more evidence was uncovered.

The 137-year-old mystery was solved in 2007 with the announcement that just such evidence had been found in Schoharie County in the form of a fossilised tree crown and a trunk with root material still attached. Analysis showed that they were fossil remains of Wattieza trees from the mid-Devonian period and were of the same type as those of the Gilboa stumps found about ten miles (16 km) from the site. They dated to 385 million years ago, a time some 140 million years before the dinosaurs and when even amphibians hadn't managed to clamber out of water.

Marks on the trunk revealed that the branches of the trees fell off as it grew taller so that they were only present close to the top of the plant. During its lifetime, each tree shed an estimated 200 branches to litter the forest floor. The Wattiezas of the Gilboa forest were the biggest lifeforms of their age, towering above other plants, and with long trunks capped by a crown they looked similar to modern tree ferns.

The fossil specimen was about 26 feet (8 m) tall, but the thickness of the base of the trunk suggests it would have been a tiddler compared to a number of other stumps at Gilboa. Some of those trees would have soared to at least 40 feet (12 m) tall, and quite possibly more.

Leaves of the Wattieza genus were more like twigs than the flat, green leaves of today's oaks and beeches, so the forest canopy would have been much thinner and let more sunlight reach the ground.

The impact Wattiezas had on the environment was enormous. Not only did they cover vast swathes of land, mainly coastal and other lowland areas, but they changed the chemical composition of the atmosphere. In the mid-Devonian period the carbon content of the air was much higher than that of today. As the trees grew and took over new territory, they absorbed and trapped carbon, bringing down the temperature and changing the balance of gases in the atmosphere to a mix similar to the present day. Wattieza trees were so successful at taking carbon out of the atmosphere that they enabled broad-leafed trees to develop, which then took over the land from them some 20 million years later.

The crown and trunk fossils were found during digs carried out in 2004 and 2005 by researchers at Binghamton University, New York, and the New York State Museum, and were identified after a world expert on Wattiezas, Dr Christopher Berry of Cardiff University, in the UK, was called in to help.

Dr Berry said the discovery of the fossils, which were preserved after being covered in sediment in a river delta, gives scientists a much better understanding of the first forest ecosystems. Through further digs, researchers hope to learn more about the creatures that inhabited the forest. A large millipede is known to have been crawling through the rotting forest litter, and many more, such as the ancestors of spiders, are assumed to have existed there.

The drowned

forest

Club mosses found on earth today grow to a maximum of about two inches (5 cm) tall, but 300 million years ago they would have dominated the landscape as 130-foot (40 m) monsters. Fossil remains of the huge mosses have been found deep beneath the surface in a network of passages carved out by coal miners in Illinois, US. They were among 50 different species of ancient plants that formed a forest 300 million years ago and were preserved as fossils at a time when reptiles had just evolved, the dominant animals were arthropods, and giant insects—such as six-foot-long woodlice and dragonflies with wingspans of two feet—roamed the earth.

The remains in the coal mines are so extensive that they form the largest fossil forest yet discovered, covering 10,000 hectares—as big as the UK city of Bristol. Preservation on such a scale was made possible by an earthquake that caused a huge segment of coastal land to sink suddenly by 15 to 30 feet (4.5 to 9 m). Much of the forest was covered immediately by water and, crucially, thick mud and many more plants sank beneath the surface over the following months. With thousands of trees and shrubs covered in mud almost simultaneously, the fossils provide a remarkable snapshot of forest life 300 million years ago during the Carboniferous era.

Paleontologists viewing the drowned forest can see it from underneath. Coal formed over millions of years from the peaty forest debris that nourished the trees and shrubs. Miners have dug out the coal, leaving the fossils visible in the ceilings of the passages they created more than 200 feet (61 m) underground. So complete was the preservation that entire toppled trees and shrubs can be seen, complete with their leaves and branches.

The range of plant species in the forest has transformed modern understanding of the Carboniferous forest landscape. Researchers were staggered by the number of trees and plants in the drowned forest. While it contained less than a tenth of the 600 species found per hectare in a modern tropical forest, there was much greater variety than previously realised.

Alongside the giant club mosses, which would have dwarfed everything else around them, were horsetails and tree ferns, themselves growing to heights of more than 60 feet (18 m), and shrubs such as pteridosperms, seed-bearing plants that are now extinct but represent the evolutionary link to today's conifers and ferns. So many fossilised plants have been found over such a wide area that researchers expect to analyse how the species range changed with the landscape and the local environment.

To see the forest was, said Dr Howard Falcon-Lang of the University of Bristol, an "amazing experience" but one that will be experienced by only a handful of people. Safety concerns meant that while he and other scientists mapped hundreds of miles of subterranean fossil-filled passageways, any attempt to photograph the features had to be accompanied by gas measurements to ensure the camera flash did not spark an explosion. Even without explosions the disused mine passages will not survive for long. The roof supports will eventually give way, probably within a decade, and millions of tons of rock will collapse into the passageways, closing them off.

It is the largest fossil forest yet discovered, covering 10,000 hectares—as big as the UK city of Bristol

opposite, clockwise from top left: Three fossilised portions of a 300 million-year-old forest—a spore-producing cone of a calamite tree, a small branch from the crown of a lycopsid tree, and leafy branches of a calamite tree
[photos courtesy of Howard Falcon-Lang, University of Bristol]

First life

Evidence of the first life on Earth has emerged from Western Australia, where researchers believe they have identified the oldest fossils in existence. The stromatolites come in a variety of shapes, including those of ice cream cones and egg cartons, and date back 3.43 billion years, about 1.1 billion years after the planet was formed.

Fresh evidence that the structures, found over a six-mile (10 km) stretch, are of biological rather than chemical origin came when a study led by Macquarie University in Sydney pinpointed seven classes of stromatolite. The origin of the ancient forms has been disputed for three decades, with scientists divided between either writing them off as having developed through the chemical processes of hydrothermal vents or accepting them as indications of microbial life.

The microbial theory was boosted by the new research, which concluded the complexity of the structures in the Strelley Pool Chert rock formation in the Pilbara region was such that they had to have been formed by the interaction of microbes and sediment in a marine environment. A resemblance to reef structures was also cited by the research team, which was convinced the stromatolites revealed not just microbial life, but diversity sufficient to represent an ecosystem. Abigail Allwood of Macquarie University said the stromatolite structure changed depending on how close to the surface of the sea they were, showing that each of the seven types had their own environmental niche. The study found that the microbial reef system had greater variety and complexity among the stromatolites, which would have had more sunlight available to them.

The stromatolites date back 3.43 billion years, about 1.1 billion years after the planet was formed

right, from top: Three examples of stromatolites found at Shark Bay, Australia—mushroomed-shaped, irregularly shaped, and large dome-shaped stromatolites [photos courtesy of Abby Allwood]

First maternal mum

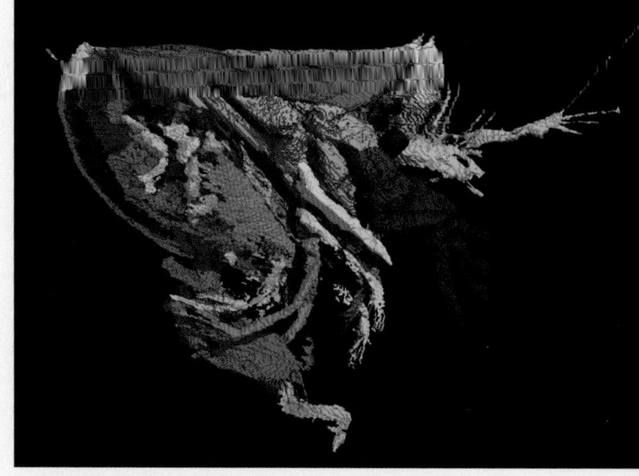

above: *Nymphatelina gravida* [illustration courtesy of David J. Siveter, Derek J. Siveter, M. D. Sutton, and D. E. G. Briggs]

The earliest example of maternal care was unearthed from volcanic rock in Hertfordshire in the UK. Inside a fossilised crustacean dating back 425 million years, 20 eggs were found, each about two one-hundredths of an inch (0.5 mm) across, as well as two juveniles. The presence of the juvenile ostracods inside the female suggests that, rather than simply lay eggs and leave the young to look after themselves, the mother continued to care for them after their birth.

So well preserved was the tiny pregnant ostracod, a relative of shrimps, that along with the eggs and young, researchers were able to identify legs and eyes. The joint UK and US team of researchers, led by Professor David Siveter of the University of Leicester, was astonished at the detail provided by the fossil.

Eggs and soft parts of the anatomy are unusual in the fossils of large creatures, but it is especially rare to find them preserved in an invertebrate. The ancient mother was named *Nymphatelina gravida*, or "pregnant young woman of the sea", and researchers said the fossil provides unique insight into parental development during the Silurian period. Embryos that are even older have been discovered, and a new imaging technique has allowed scientists to examine 500 million-year-old specimens in three dimensions for the first time. The embryos from an ancient worm-like species represent some of the earliest multi-cellular life on the planet, and 3D images were produced during research led by the University of Bristol in the UK after using synchrotron radiation X-ray tomography. Development of the embryos can be seen through different stages from the point when cells started dividing, and the technique allows tiny differences between closely related species to be observed and analysed.

Horned crustacean

An ancestor of lobsters that had more than 100 legs and six horns on its head has been discovered at the same site as the maternal crustacean. The bizarre new creature, *Tanazios dokeron*, appears to be a link between crustaceans and insects. Professor Siveter, a member of the team that analysed the find, described it as "a real weirdo". At just over an inch (2.5 cm) long, it is thought to have been a scavenger living at the bottom of a tropical sea 425 million years ago. Researchers believe that, despite being blind, it was in all probability an active swimmer.

The animal was a precursor to crustaceans such as crabs and shrimps and had two flaps—called epipodites—that on insects eventually evolved into wings. The finding supports genetic studies that have indicated a close relationship between insects and crustaceans.

Researchers were unable to cut the fossil free of the calcium carbonate nodule it was encased in, so they used state-of-the-art technology to uncover the animal's structure. The surface of the rock was ground away in slices 20 microns (one-fiftieth of a millimetre) thick and digital pictures were taken with each cut. Though it meant the destruction of the fossil, by putting all the digital photographs together, the research team was able to see the mysterious creature in three dimensions and in enormous detail.

Furry glider

below: Illustration of *Volaticotherium antiquus* [courtesy of Zhao Chuang and Xing Lida]

A gliding squirrel has pushed back the date of the first mammalian flight by more than 70 million years after being unearthed in Mongolia. Until the discovery of *Volaticotherium antiquus*, meaning "ancient gliding beast", the earliest known mammal to take to the air was a bat from 51 million years ago. *V. antiquus*, however, was gliding between the trees 125 million years ago, a time when even the ancestors of birds were still trying to master the art of flying.

Such early evidence of flight among mammals suggests, said the scientists who analysed the remains, that early mammals were more diverse than previously believed. Remains of the animal, a cross between a bat and a squirrel, were so well preserved that impressions of fur and skin were found on the rock that contained the fossil. It had sharp teeth similar to those of bats and would have used them to munch on insects, though it is unlikely to have been a good enough flier to catch them in the air. Its limb structure was suited to running up trees, from where it would have launched itself into flight. Skin stretched between limbs, and its low body weight of about 2.5 ounces (70 g) meant it would have been able to glide from one branch to another while a long, stiff tail helped control the direction of the flight.

Analysis of the fossil showed that the creature came from a previously unknown order of mammals, meaning that it developed flight independently from the ancestors of modern flying squirrels, bats, and other flying mammals. The research team, from the American Museum of Natural History in New York, US, and the Chinese Academy of Science in Beijing, said the dating put the mammalian glider in the Cretaceous forests at the same time as proto-birds were taking off.

The mammal was gliding between the trees 125 million years ago at a time when even the ancestors of birds were still mastering the art of flying

Prehistoric Jaws

An ancient sea monster has been found to have had the most powerful bite yet identified in the history of evolution. Its teeth could snap together with such power that it could have torn chunks out of anything in its range.

Such pressure was exerted when it bit down that a bite from Jaws, the great white shark that terrified a generation of cinemagoers, would be a nibble by comparison. *Dunkleosteus terrelli* was able to clamp its front teeth with a pressure of 80,000 pounds per square inch (5,600 kg per sq cm).

But power was not its only advantage. It could open its mouth so fast—in just a fiftieth of a second—that prey would have been sucked straight in. Even *Tyrannosaurus Rex*, an icon among predatory dinosaurs, couldn't have competed. Its bite has been calculated to have exerted 3,000 pounds per square inch (1,360 kg per sq cm), about a quarter of the marine monster's.

Of animals around today, the American alligator has the most powerful bite, with a force of 2,125 pounds (963 kg). Humans muster 170 pounds (77 kg), though a 10-stone (63 kg) woman exerts 279 pounds (127 kg) through a stiletto heel.

The armoured sea creature lived about 400 million years ago in the Devonian period and grew up to 33 feet (10 m) long, weighing in at four tonnes. It belonged to a class of fish known as placoderms and might have eaten early sharks, which first developed in the Devonian period. Its huge bite would have enabled it to gnaw through a shark with a single snap of its jaws. Nevertheless, such extraordinary power was no guarantee of survival, and competition from the smaller but faster and more agile shark is cited as a probable cause of *D. terrelli's* ultimate extinction.

To calculate the power of the creature's bite, researchers from the Field Museum in Chicago, US, used a fossilised skull to work out where the muscles would have sat and what size they would have been. They discovered that it boasted four rotational joints in its jaw, which were responsible for the speed with which it could open and close its mouth.

A bite from Jaws would be a nibble by comparison

below: *Dunkleosteus terrelli* head and dorsal trunk plate [photo: © Natural History Museum, London]

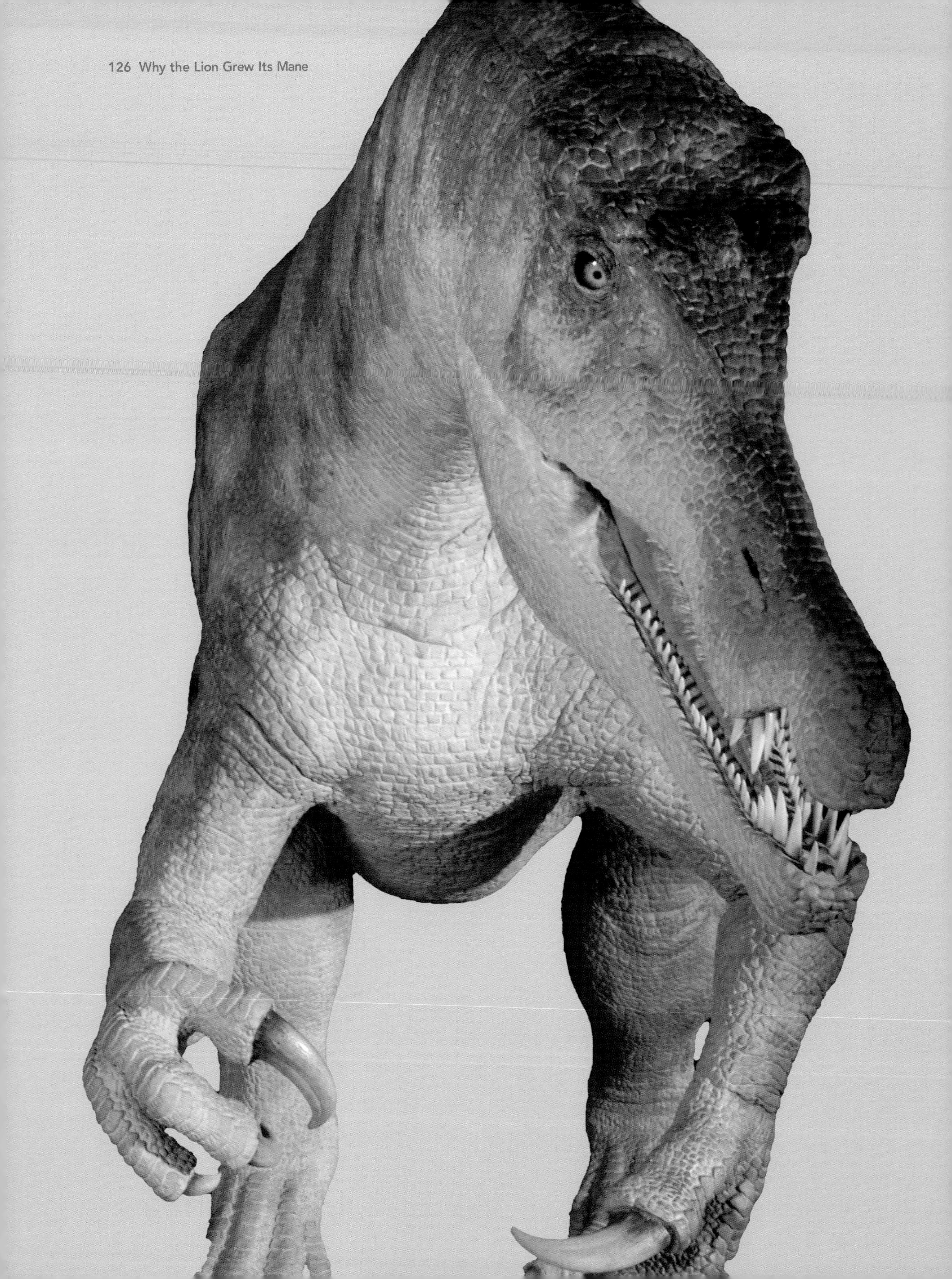

Euro-monster

One of the biggest animals ever to walk the earth was identified in 2006 by a team of paleontologists in Spain. Named *Turiasaurus riodevensis*, it is estimated to have stretched 100 to 120 feet (30 to 37 m) from the tip of its nose to the tip of its tail, and weighed 40 to 48 tonnes. It was rivalled in size only by other long-necked sauropods, such as *Argentinosaurus*, a 120-foot (36 m) creature from South America that weighed 100 tonnes, and Africa's *Paralititan stromeri*, which was 100 feet (30 m) long and weighed 75 tonnes.

While *Argentinosaurus* maintained its position as the world's biggest creature, *T. riodevensis* claimed the place of Europe's largest-ever land animal when it was unearthed. It lived about 150 million years ago and was herbivorous. One of its leg bones, the humerus, was as tall as a human and its weight made it a match for at least six modern-day elephants.

T. riodevensis was found close to the village of Riodeva in the Teruel region, where the rock formations are regarded as among the best in Europe for fossil hunters. Dr Luis Alcalá of the Dinopolis Museum in Teruel led the dig, which started in 2003, and said so many fossilised bones from the animal had been found that it was possible to identify it as part of a new group, not just a new dinosaur species. Analysis showed that its teeth were similar to many other fossils dug up in other parts of Europe, including France, the UK and Portugal, suggesting that it roamed over much of the continent.

Similarly, analysis of the teeth of *Baryonyx*, a fish-eating dinosaur first dug up in Surrey, England, suggests a more widespread habitat than initially suspected. Its discovery in 1983 has forced a recent reassessment of many of the ancient crocodile teeth that have been unearthed over the last 150 years. Many of the teeth attributed to crocodiles were found to be identical to *Baryonyx*, and about one in 10 of those housed at the Natural History Museum in London have now been re-identified as belonging to the fish-eater.

Baryonyx, which lived about 125 million years ago, has forced more than just a reassessment of fossilised teeth. A specimen found in Thailand has required scientists to reconsider when Asia and Europe were joined.

The creature, which lived by rivers, lakes and the sea where it could scoop up fish from shallow waters, may still prove to have spawned the greatest meat-eaters as well. A seven-foot-long skull found in Africa was identified as

Baryonyx specimens found in the UK and Thailand have forced scientists to reconsider when Asia and Europe were joined

belonging to *Spinosaurus*, an immediate descendent of *Baryonyx*, but as none of the body was found it was impossible to judge the overall size of the animal. However, with such a large skull it has the potential to be bigger than *Gigantosaurus*—which, with a six-foot skull, is accepted as the largest meat-eater yet identified.

below and opposite: An animatronic model of *Baryonyx* [photos: © Natural History Museum, London]

An animatronic model of *Baryonyx*
[photo: © Natural History Museum, London]

Digging dino

The discovery of an adult and two juveniles, still in a hole 95 million years after their deaths, provided the first solid evidence that dinosaurs could burrow. Sand had filled the twisting burrow, which was more than six feet (2 m) long, after the animals died, and over time it turned to sandstone. Because it cut through three distinct layers of rock and contained the fossilised bones of three animals, researchers were quickly convinced the sandstone represented a dinosaur burrow.

Dr David Varricchio of Montana State University in the US led the study and said the finding meant that small dinosaurs would have had a greater range of territories available to them because of the protection from the elements provided by a burrow. Among the regions where a burrow would have made it possible for a small dinosaur to live are high altitudes, where it would have helped keep them warm, and deserts, where it would have protected them from extreme heat. Of course, it would have provided shelter from predators in any terrain. The fossil burrow was found in an old river floodplain in Montana and, at approximately 16 inches (40 cm) high and 12 inches (30 cm) wide, was about the ideal size for the dinosaur—big enough to crawl through but too small for larger creatures to enter.

Each of the dinosaurs found at the bottom of the burrow were of a previously unknown species, now named *Oryctodromeus cubicularis*, which translates as "digging runner of the lair". Analysis of the bone structure confirmed that the dinosaurs shared body characteristics similar to modern digging animals, making it likely that they did not die in a hole made by another creature. It would have walked on two legs, but had strong shoulders and forearms adapted to dig, and perhaps was able to use its snout to help push dirt out of the way.

Small dinosaurs would have had a greater range of territories available to them, with burrows providing protection from both predators and the elements

That the adult was found with two juveniles provided strong additional evidence to support the idea that at least some dinosaurs cared for their young after they hatched, just as do modern birds. The adult in the burrow measured about seven feet (2.1 m)—its tail comprising more than half its length—and is estimated to have weighed 48 to 70 pounds (22 to 32 kg), making it about the same size as a modern coyote. The juveniles were about half the length of the parent.

Finding the animal remains within the hole provided convincing evidence that *O. cubicularis* burrowed and may help explain other finds. When studying a related dinosaur called *Orodromeus*, Dr Varricchio came across a couple of instances where bones from a floodplain environment had stayed together. Had they been on the surface, they would have been expected to scatter and separate, so being found together may indicate they had been in a burrow.

above: Illustration of a *Oryctodromeus cubicularis* head
below: Illustration of an adult *O. cubicularis*, showing the recovered bones in their expected place, and silhouette of a juvenile in grey
[images courtesy of Lee Hall, Montana State University]

Early bird catches the fish

Ancestors of modern birds relied more on swimming than flying to survive

Fossils from 110 million years ago suggest the ancestors of modern birds relied more on swimming than flying to survive. Studies of five fossils, found near Changma in the Gansu region of China, show the early bird swam like a duck and dived like a grebe—and would have caught a fish rather than a worm.

Gansus yumenensis was a form of waterfowl that is thought to be the oldest member of the Ornithurae lineage, which includes modern birds. It split from other early birds, of the Enantiornithes group, which differed in that they had reversed shoulder joints. These "opposite birds" left no descendants when they died out 65 million years ago. Apart from the fossils' missing heads, the state of preservation is so good that feathers and even the webbing on a foot could be detected.

Hai-lu You, of the Chinese Academy of Geological Sciences, led the research and concluded that the creature was aquatic and had a similar lifestyle to modern birds such as ducks, herons and loons. Webbing and the structure of the legs suggested the early bird frequently dived, was a strong swimmer, and would have been able to take off from the surface of the lakes it lived on. Professor Peter Dobson of the University of Pennsylvania in the United States helped analyse the fossils as part of the Chinese-American research and said the remains revealed *Gansus* to be the oldest known "nearly-modern" bird.

right: Illustration of *Gansus yumenensis*
above right: Skeletal reconstruction of *G. yumenensis*, with the grey-shaded bones approximated using closely related fossil birds
[illustrations courtesy of Mark A. Klingler / CMNH]

Waspish bee

An insect that died 100 million years ago has been described as the missing link between wasps and bees after it was found preserved in amber. Its discovery supports the theory that bees evolved into today's plant pollinators from an ancient meat-eating wasp.

The creature appeared to be a cross between the two insects, but was declared to be the world's oldest bee after careful assessment of its structure. It has some body parts typical of wasps, but also branched hairs, presumed to be for pollen collection, and other features associated with bees. It predates the next oldest-known bee by at least 35 million years, and was found encased in amber in a mine in the Hukawng Valley in Myanmar. The male bee, called *Melittosphex burmensis*, died when it got stuck in tree resin and was preserved in such perfect detail that individual hairs can be identified. A wing, legs, thorax, head and abdomen are all clearly visible.

Bees evolved into today's plant pollinators from an ancient meat-eating wasp

Pollen-bearing flowers only evolved about 125 million years ago, and by the time the bee was flying it took over as the dominant means of plant reproduction from the windborne pollination of conifers. The bee is only 12 one-hundredths of an inch (3 mm) long, but its size is consistent with the smallness of many of the flowers produced by plants 100 million years ago.

Professor George Poinar of Oregon State University in the US led the team that analysed the amber insect and concluded it was more bee than wasp. He said that the amber insect is crucial to the understanding of bee evolution. The shared bee and wasp features of the specimen back up the idea that the two insects have a common ancestor, and researchers hope the amber insect will help identify the point at which they split and provide insights into the sudden spread of flowers during the Cretaceous.

above right: ***Melittosphex burmensis*** **encased in amber [photo courtesy of George Poinar]**

Amber spider

A spider trapped in amber 115 to 121 million years ago put back the date at which they were spinning the orb webs common in today's gardens by more than 20 million years. Until the specimen was identified by Dr David Penney of the University of Manchester in the UK, the earliest they were known to have existed was 94 million years ago.

Like the early bee, the spider was preserved in remarkable detail and its eyes, tarsal claw structure and reproductive organs could be identified. Dr Penney, who found the amber in a collection held by the Museo de Ciencias Naturales de Álava in Vitoria, Spain, said it showed all the major orb-weaving families of spiders had evolved by the Lower Cretaceous.

He observed that it meant the orb web, used mainly to catch flying insects, was already in use at a time before many of the plant-pollinating insects existed. The spider came from the *Araneidae* and is the first known to have been living during the Lower Cretaceous.

The orb web, used mainly to catch flying insects, was already in use at a time before many of the plant-pollinating insects existed

below: The spider, measuring only eight one-hundredths of an inch (2 mm), preserved in amber [photo: David Penney]

Biplane bird

Some 125 million years before the Wright brothers took off from the ground in the first biplane in 1903, four-winged birds had already taken up the design. Fossils of *Microraptor gui* were first found in 2000 with an unusual arrangement of feathers noticed on its hind legs. Initially it was thought the leg feathers lined up behind the main wings to create a second set, much like the four-wing arrangement in dragonflies, but in 2007 fresh analysis by scientists at Texas Tech University in the US concluded the bird was designed like a biplane.

Like a biplane with a staggered array of wings, the bird would have been adept at gliding between trees

By studying the aerodynamics of the proto-bird, they found that too much turbulence would have been caused if the wings were arranged in a dragonfly-fashion. Instead, the researchers calculated that with the leg wings placed behind and below the main wings, like a biplane with a staggered array of wings, the bird would have been much more adept at gliding between trees. *Microraptor gui*, which weighed about two pounds (1 kg) and had a wingspan of approximately three feet (1 m), would have been unable to take off from the ground, even with a run-up, but it would have been able to glide from branches.

The fossils, from China, support the idea that flying developed through animals making use of gravity to glide downwards, instead of the alternative theory that they evolved into fliers from the ground by running, hopping and flapping. Biplane designs in birds could simply have proved to be an evolutionary experiment that failed, but it could help explain the leg feathers found in modern birds of prey. Feathers on legs help streamline birds of prey as they make their aerial attacks. A gradual shift from biplane to two-winged designs, while retaining some of the aerodynamic qualities of leg feathers, is a theory supported by other fossil remains.

above: Illustration of *Microraptor gui* [© Andrey Atuchin / NHM]

Jurassic beaver

A beaver that lived almost 100 million years before the dinosaurs went extinct has been identified as the earliest mammal to live an aquatic lifestyle. *Castorocauda lutrasimilis,* meaning "otter-like beaver-tail", grew to at least 17 inches (42.5 cm) long and was the biggest mammal ever to have lived when it was catching fish 164 million years ago.

It was the biggest mammal ever to have lived when it was catching fish 164 million years ago

Its fossilised remains were unearthed in China and it is one of several discoveries that suggest that the mammals of the Jurassic period were more diverse than previously suspected. Until recently, it was thought that mammals of the period were limited to insect-eating rat and shrew-like creatures.

Dr Quiang Ji of the Chinese Academy of Geological Sciences in Beijing, who led the team that discovered the fossil, said the fish-eating animal was capable of paddling and remnants of soft tissue indicated it had webbed hind feet. Other firsts for mammals garnered by the creature from the Jurassic period were fur and scales, which were seen as imprints in the fossil. Powerful limbs hinted at its ability to swim, while the seal-like teeth pointed to a diet of fish. Although its skeleton is similar to modern otters, its flat tail instantly brings to mind modern beavers.

6 cm

right: *Castorocauda lutrasimilis* skeleton reconstruction
below: *C. lutrasimilis* fur diagram
[illustrations courtesy of Quiang Ji, © Mark A. Klingler / CMNH]

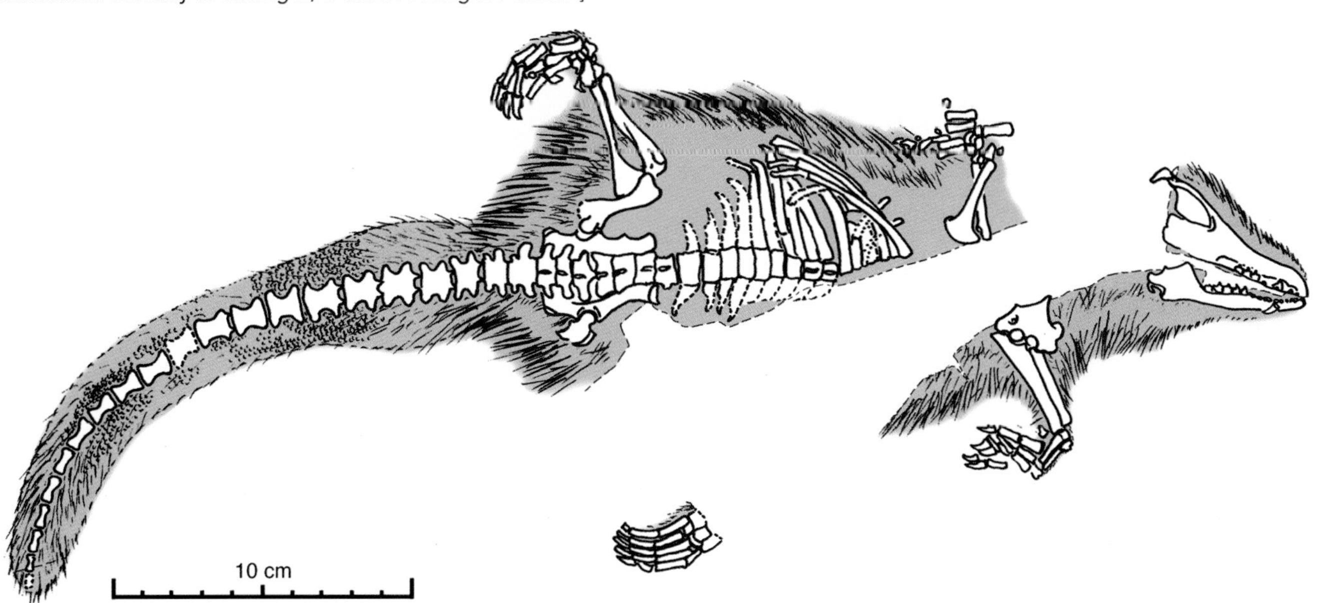

Fishing link

A predator with a head like a crocodile but gills like a fish has been identified as perhaps the first animal to be able to do push-ups. *Tiktaalik roseae* is a missing link in the evolutionary lineage between finned fish and limbed land animals, or tetrapods.

The so-called fishapod is thought to have lived 375 million years ago in the shallow and slow-moving waters of a river delta system and may have been able to clamber out onto land for short periods. Its front limbs were still fins, but within them had developed wrists and shoulders that allowed the animal to push up its head and chest and support its own weight. The fossil remains were found in the Canadian Arctic in rocks that were once in a sub-tropical region during the late Devonian period 380 to 365 million years ago.

Paleontologists had realised a creature of that type must have existed because of the gaps in the fossil record between fish and land animals. A US team led by Dr Neil Shubin of the University of Chicago, Dr Edward Daeschler of the Academy of Natural Sciences in Philadelphia, and Dr Farish A. Jenkins, Jr of Harvard University set out to find it after fixing on Ellesmere Island as a likely place to find the evidence.

The creature's ability to support itself on land was a key factor in identifying it as an evolutionary halfway house between fish and the limbed creatures that eventually evolved into humans. Strong ribs are essential for the sturdy trunk required by land animals. Fish, which are supported by the water, manage with weaker ribs.

Tiktaalik roseae, however, had developed a more robust ribcage, similar to that of an anteater.

Other features that pointed to the animal representing the structural shift required to move from water to land included the flat head reminiscent of a crocodile, while retaining the scales of a fish. Equally, it is the only fish to possess a neck, which meant that with the skull separated from the shoulder girdle it would have more freedom of movement.

The creature was an evolutionary halfway house between fish and the limbed creatures that eventually evolved into humans

below left: Illustration of *Tiktaalik* and its fossilised remains
below right: Diagram illustrating how *Tiktaalik roseae* is a transitional species between lobe-finned fish and tetrapods of the Devonian period [illustrations: Kalliopi Monoyios]

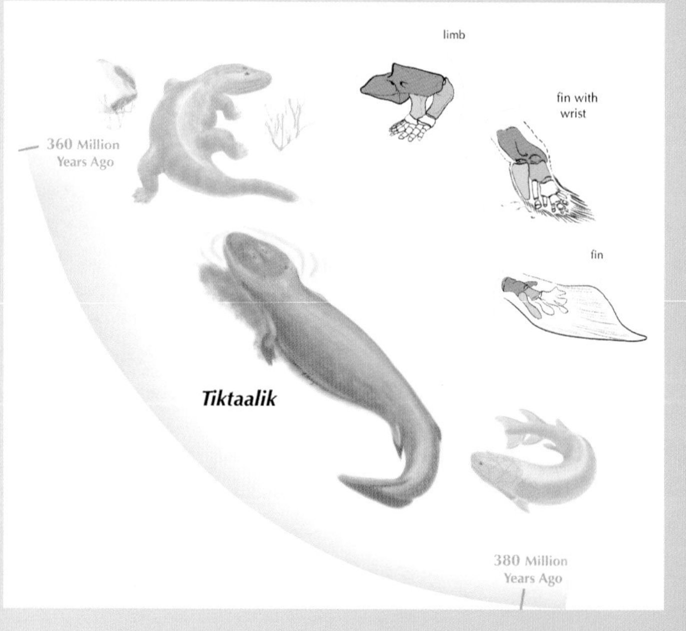

Shrinking dinosaur

As the seas rose to create islands, the huge creatures had to adapt by getting smaller

A dinosaur that measured 20 feet (6.2 m) long was a mere midget compared to its relatives, and provides perhaps the greatest example of evolutionary shrinkage.

Sauropods were the biggest animals ever to walk the earth, but the smallest member of the family, *Europasaurus holgeri*, weighed little more than a third of a modern elephant. By comparison, the closely related Brachiosaurus grew to 279 feet (85 m) and weighed up to 80 tonnes, whereas *Europasaurus* was a lightweight one tonne.

The ancestors of the *Europasaurus* would have been much bigger, but they became trapped in a slowly shrinking territory in the Lower Saxony basin in Germany. Their territory would have been islands in the region which, along with much of Central Europe, was largely submerged by the sea.

As the seas rose to create the islands, the largest of which would eventually have been about 77,200 square miles (200,000 sq km) in area, the huge creatures had to adapt by getting smaller—otherwise the territory was simply too small to support them.

A study of fossil bones led by Dr Martin Sander of the University of Bonn revealed that the 20-foot specimen of *Europasaurus* from 150 million years ago was not a juvenile as initially assumed, but instead a fully grown adult.

above: *Europasaurus holgeri* [illustration © Gerhard Boeggemann, used under the Creative Commons License]

previous, clockwise from top left: Computer model of a double-stranded RNA molecule [image: © Laguna Design / Science Photo Library], coloured scanning electron micrograph of *C. difficile* bacteria [image: © D. Phillips / Science Photo Library], Irish wolfhound and a border terrier [photo courtesy of Kenneth Sutter], Californian purple sea urchins [photo courtesy of Alex Lin, Eileen Fong, Jin-Hong Kim / California Institute of Technology], Rafflesia flower in Mount Kinabalu National Park, Borneo [photo: © NHPA / Mark Bowler], apples in an array of colours [photo courtesy of Scott Bauer]

All in the Genes

Giant leaps in understanding the roles and identities of genes in humans and other animals have been made in little more than a generation. Biochemist Dr Fred Sanger, the two-time Nobel laureate, and colleagues first developed rapid genetic sequencing in 1975—a technique that is still in use, making possible much of the genetic research carried out today.

Other genetic landmarks have been reached since then, not least the DNA fingerprinting technique pioneered by Professor Sir Alec Jeffreys that has proved such an important crime-fighting tool. Equally, the genome—the collection of genetic material contained by a species—has been successfully mapped for humans and an increasing number of animals.

Genetics has grown in just a few decades from a small area of study into an enormous research field that has an increasingly pervasive influence on all our lives, whether in finding cures for disease, saving animals from extinction or simply helping people trace their ancestry.

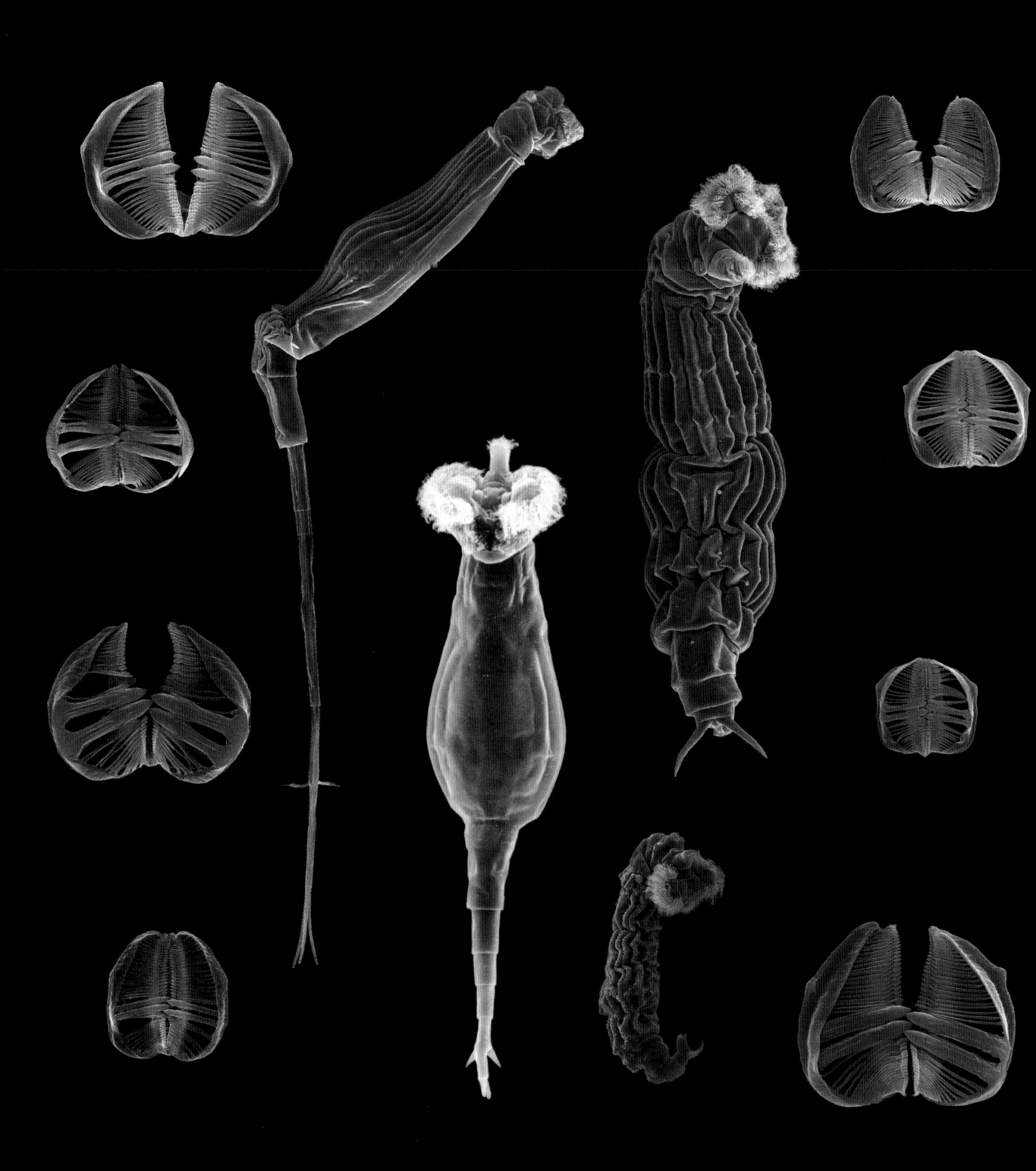

Sexless evolution

Bdelloid rotifers have managed to evolve into almost 400 different species despite only reproducing asexually

A creature that has gone without sex for an estimated 100 million years or more has disproved the notion that asexual reproduction is an evolutionary dead end. Bdelloid rotifers, microscopic animals little more than three times the length of a human sperm, have managed to evolve into almost 400 different species despite only reproducing asexually.

The diversification of bdelloid rotifers into different species was shown through genetic analysis, combined with measurements of jaw sizes made possible by a scanning electron microscope. DNA sequencing was used to build up a profile of various types of the aquatic creature, showing that they differed sufficiently to be considered separate species. Furthermore, the different types had evolved to take advantage of a variety of ecological niches in the rivers, ponds, soils, lichens and mosses that they inhabit. It was previously argued that variations between bdelloid rotifers had emerged by chance through random mutations, but the new study concluded they evolved through natural selection.

Asexual reproduction is usually assumed to condemn a species to extinction after about a million years. Yet the bdelloid rotifers are known from fossil records to have been around for 40 million years without sexually reproducing, while DNA analysis suggests they go back at least 100 million years. It is thought that to survive, they have mastered the art of divergent selection—the process by which life-forms diverge into separate species to take advantage of differing environments. Asexual life-forms usually alter structure, as they clone imperfectly at the cost of their original form.

Dr Tim Barraclough of Imperial College London in the UK, who worked on the project with Dr Diego Fontaneto of the University of Milan, Italy, said one of the most striking illustrations of the creatures' ability to evolve was found on a water louse. One type of bdelloid rotifer lived around the legs, while a second lived on the chest—they arrived on the louse as one species but then diverged to take fuller advantage of each niche.

opposite: Scanning electron micrographs showing variations of bdelloid rotifers and their jaws
overleaf: Scanning electron micrographs of hard jaws, or trophi, of bdelloid rotifers [images courtesy of Diego Fontaneto]

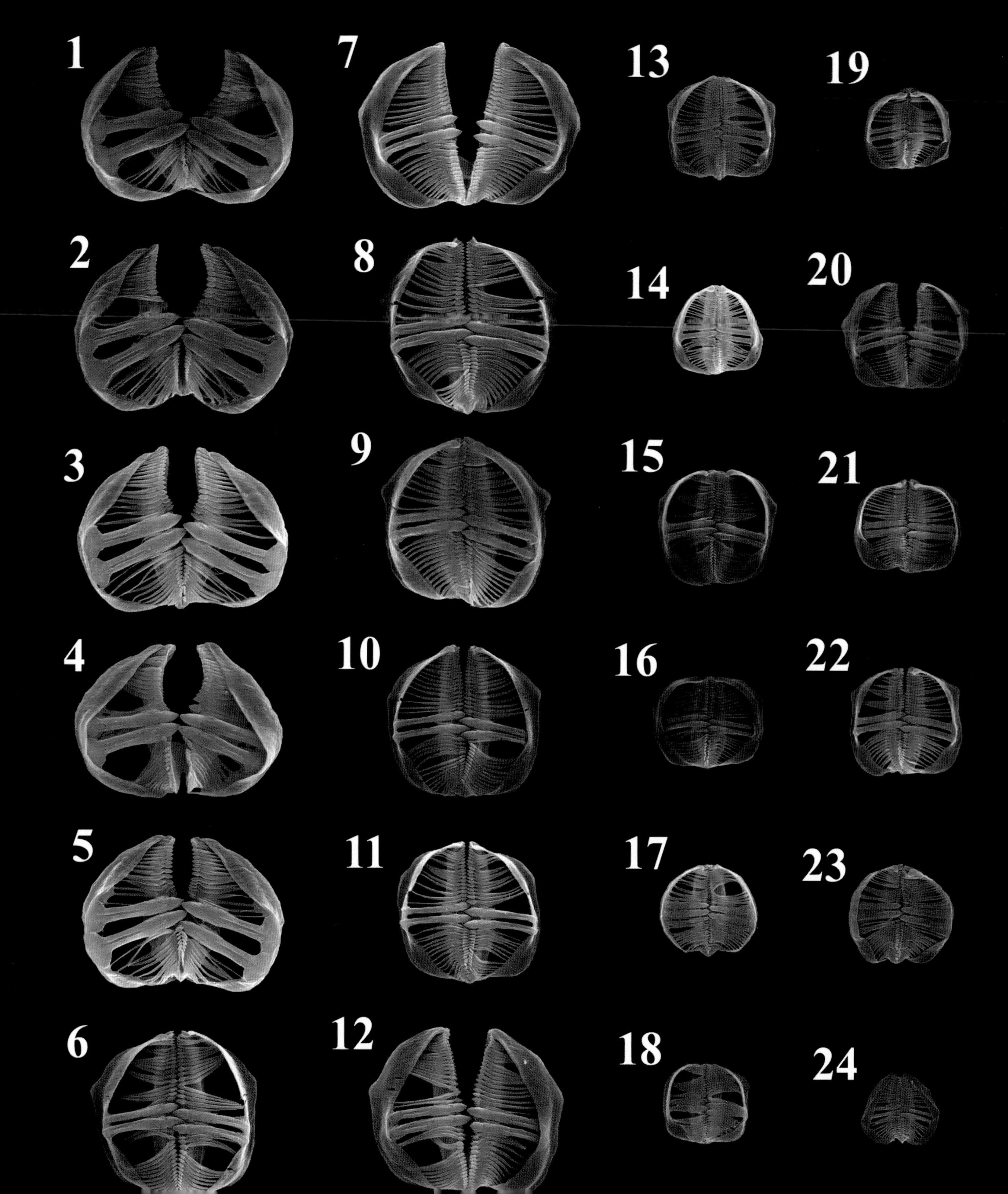
1
7
13
19
2
8
14
20
3
9
15
21
4
10
16
22
5
11
17
23
6
12
18
24

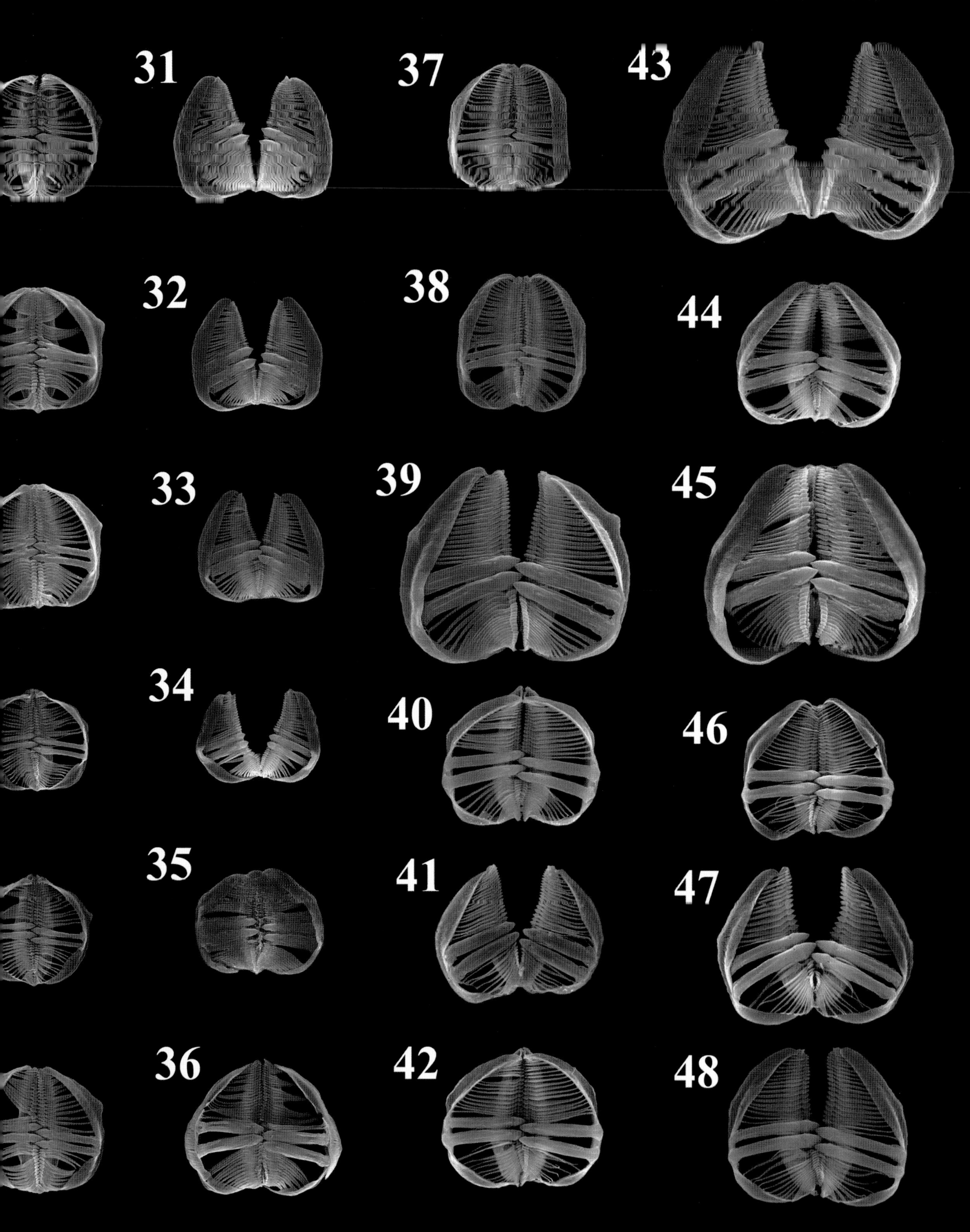
31
32
33
34
35
36
37
38
39
40
41
42
43
44
45
46
47
48

And then there was one

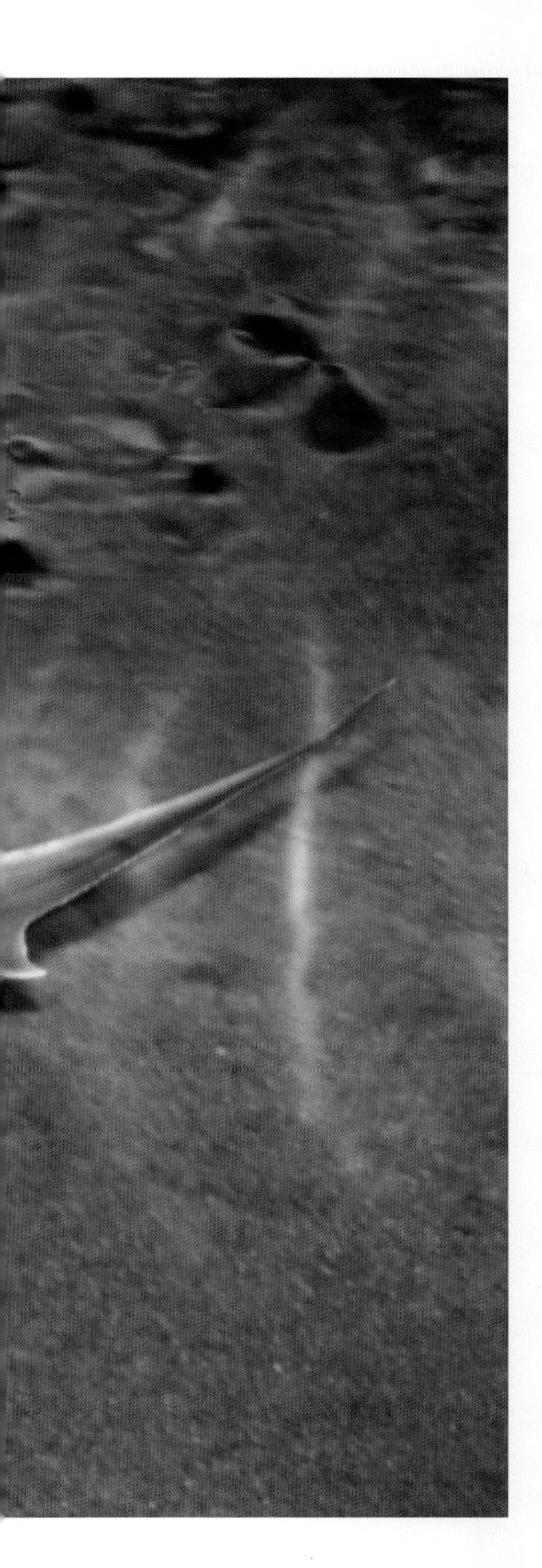

The pup, whose mother had gone three years without contact with a male shark, lacked any trace of paternal DNA

DNA profiling of three female sharks and the remains of a baby shark has revealed that fish can breed parthenogenetically. The finding means that of all the major vertebrate animal groups on the planet, the only creatures unable to reproduce asexually are mammals.

The bonnethead shark, *Sphyrna tiburo*, a type of hammerhead, was born at the Henry Doorly Zoo in the US in 2001. While several cases of suspected parthenogenetic shark births had been reported anecdotally, none were confirmed until 2007 when the DNA of the Henry Doorly bonnetheads was analysed. The results of the DNA barcoding revealed that the pup—or at least its remains, since it was killed by a stingray—lacked any trace of paternal DNA. Researchers were able to identify the mother shark because it shared significantly more DNA with the pup than the other two adults in the tank.

Previous cases where pups were born in the absence of male sharks were attributed to delayed fertilisation, where sperm is stored in the mother's body after mating instead of immediately fertilising the egg. With the Henry Doorly bonnetheads, there were full records showing that the adult females had been in captivity and out of contact with males for at least three years. All three were taken from the wild before reaching sexual maturity and, regardless, previous research suggested the maximum time sperm could be stored in a female's body was five months.

Until the study, led by researchers from the Guy Harvey Research Institute in the US and Queen's University, Belfast in Northern Ireland, proof that fish could breed parthenogenetically had remained elusive. Genetic barcoding provided the evidence, and Dr Mahmood Shivji of the Guy Harvey Research Institute said parthenogenesis was "the most likely" explanation of other suspected fatherless shark births in aquariums.

The finding does, however, raise serious doubts about the worth of the ability to reproduce asexually in a world where the sharks are under threat. Not only did the pup miss out on the genetic variety that would have been passed down from a father, but much of the mother's genetic inheritance was absent too—a form of asexual reproduction known as automictic parthenogenesis. It was calculated by the team of researchers that the pup inherited only about half of its mother's genetic variation. The asexual breeding technique might in the short term be useful to female sharks in an environment where males are rare or nonexistent, but in the long term it would in all likeliness be disastrous because of the erosion of genetic diversity.

left: Bonnethead shark
[photo: © NHPA / Trevor McDonald]

The wrong leeches

For years medical researchers have been using one leech when they thought they had another

[illegible] valuable identification tool where traditional observations fail taxonomists. Medicinal leeches, each with three jaws lined with more than [illegible] difficult to catalogue and describe—so much so that for years medical researchers have been using one leech when they thought they had another. Several attempts were made in the years after Swedish botanist and zoologist Carolus Linnaeus first described the European medicinal leech, *Hirudo medicinalis*, in 1758, but none were entirely satisfactory.

During the nineteenth century when leeches were prescribed with abandon, *H. medicinalis* was at times considered to be up to seven different species, but problems in defining the differences meant it ended up being thought of as one. In recent decades, medicinal leeches, which have natural anaesthetics and anticoagulants in their bites, have proved a rewarding subject for study—at least 115 chemical compounds, many of them now used as valuable ingredients in drugs, have been developed as a result. Most notably, they have been used in anticoagulants such as hirudin, and antistatins that prevent viral infections including HIV and hepatitis C. Leeches themselves, once a medical mainstay with millions of them used in the hospitals of London and Paris in the nineteenth century, are even coming back into fashion as a form of treatment in plastic and reconstructive surgery.

Over the last decade, leech specialists have realised that the European medicinal leech should be considered to be at least three species, with *H. medicinalis* joined by *H. verbana* and *H. orientalis*, and probably more. The first realisation that European leeches were more than they seemed came in 1999 [illegible] taxonomist, and Dr [illegible] of the Senckenberg Museum in Germany, re-identified *H. verbana*, which was originally described in [illegible] research five years later, led by Dr Peter Trontelj from the University of Ljubljana in Slovenia. Further work on leech genes by Dr Trontelj, with Dr Serge Utevsky from V.N. Karazin Kharkiv National University, Ukraine, concluded that the *Hirudo* genus comprised five species. These findings, however, went unnoticed by many scientists and it was not until 2007 that it was found that medical researchers had been using *H. verbana* rather than, as had been widely assumed, *H. medicinalis*.

Research led by Dr Mark Siddall of the American Museum of Natural History, supported by Dr Trontelj and Dr Utevsky among others, confirmed clear differences in *H. verbana* genes compared to those of *H. medicinalis*. Equally, *H. orientalis* was considered to be distinct. The leeches analysed for the research were collected from the wild in France, Italy, Slovenia, Croatia, Macedonia, Germany, Azerbaijan and Ukraine, as well as from commercial producers.

The differences between *H. medicinalis* and *H. verbana* raised the question of which leech species has been the source of the variety of chemicals used in drug development. European medicinal leeches used in research over the last three decades have largely been accepted as *H. medicinalis*, but when researchers looked at commercial specimens sold as captive-bred leeches, the tests showed they were in fact *H. verbana*.

Both species are medicinal and both are originally from Europe, but the conclusion they are separate species raises doubts about their [illegible] instance, *H. medicinalis* was specifically named when the Food and Drug Administration approved the use of leeches for medical use to help restore blood flow after surgery. The use of *H. verbana* would seem, strictly speaking, to breach the regulations despite it being the species available commercially. Similarly, it raises doubts about the development of a range of drugs, because they would have been researched on the premise that *H. medicinalis* was used in their derivation. Further research may be required to determine whether the effectiveness of chemicals derived from leeches differs through the use of *H. verbana* instead of *H. medicinalis*, or vice versa.

In the long run, the fact that the European medicinal leech comprises more than one species is likely to lead to benefits, because it increases the range of properties that can be investigated for medicinal and industrial purposes, particularly the differences in amino acids found in the separate species. Changes in conservation rules could be necessary because while *H. medicinalis* has legal protection, *H. verbana* has none, and unscrupulous pharmaceutical companies might decide to collect them wholesale from the wild, which would threaten a population crash.

opposite: Medicinal leeches, *Hirudo medicinalis*, or is it *Hirudo verbana*? [photo: © Louise Murray / Science Photo Library]

DNA and the corpse flower

Just as DNA analysis can reveal unseen aspects of animal life, so it can be applied in botany to uncover the family trees of plants. Rafflesia, parasitic plants from the *Rafflesia* genus that boasts the biggest flower in the world, remained a mystery for almost 200 years among botanists trying to identify its near relations. Its relatives were only pinpointed when researchers from Harvard University in the US analysed its DNA and that of a variety of other plants. Genetic analysis was complicated by the fact that rafflesia had not only lost all its leaves and roots, but had adopted some of the DNA of the tropical vine it uses as a host.

Careful comparisons revealed that rafflesia is most closely related to the Euphorbiaceae family of plants, many of which produce tiny flowers hundreds of times smaller than its own. It evolved from a common ancestor 46 million years ago, over which period its flowers increased in size 79-fold. Among the plants now known to be closely related are poinsettias, Irish bells, rubber trees, and castor oil plants.

Though the family mystery appears to have been solved, rafflesia is still an enigmatic plant surrounded by unanswered questions, not least how and why it came to be so big. Rafflesia is unusual in several ways and when it was first seen by a Western naturalist it was regarded with incredulity. Joseph Arnold, who encountered the plant on Sumatra with its namesake Stamford Raffles in 1818, described it as "the greatest prodigy of the vegetable world", but harboured initial doubts about telling anyone for fear of being disbelieved.

Dr Charles Davis, who led the genetic study, said it is hard to know where to start when considering rafflesia's peculiarities. It lives entirely within a vine except when flowering. Roots, leaves, stems and photosynthesis have all been dispensed with, and it relies on the host for all nutrients. Only the flower gives its presence away to outsiders—at more than three feet (1 m) across and weighing 15 pounds (7 kg), it is the world's biggest. Even as a bud it is as big as a football. As if to further underscore its extraordinary form, it has been dubbed the corpse flower because it smells like decaying meat and emits the stench of rotting flesh to attract the carrion flies that pollinate it.

Roots, leaves, stems and photo-synthesis have all been dispensed with—the plant relies on its host for all nutrients

opposite: Rafflesia flower in Mount Kinabalu National Park, Borneo [photo: © NHPA / Mark Bowler]
above: Rafflesia drawing from page 378 of *Pflanzenleben: Erster Band: Der Bau und die Eigenschaften der Pflanzen*, Anton Kerner von Marilaun, Adolf Hansen (1913)

The colour of apples

The master gene that activates production of colour in apples has been identified and could help growers create even redder fruit. Researchers from the Commonwealth Scientific and Industrial Research Organisation (CSIRO) in Australia isolated the gene from the Cripps Pink variety and measured how active it was in producing colour in apples. They realised that the more MdMYB1 genes that were activated, the greater the intensity of the red colouration in the skins.

Light prompts the gene to activate the production of anthocyanins, natural plant compounds responsible for blue and red in many types of fruits and flowers, which turn the skin red. Activity levels of the gene are dependent on how much sunlight hits the apples.

Researchers hope that now that the master switch for colour has been identified, it will be possible for growers to exert greater control over the colour of new varieties—an important selling point on supermarket shelves. By determining the version of the gene even in seedling plants, they should be able to predict how red the eventual fruit will be. Being able to select potential varieties years before the first fruits appear would significantly reduce costs and shorten the development of new types of apples.

Growers will be able to exert greater control over colour, an important selling point on supermarket shelves

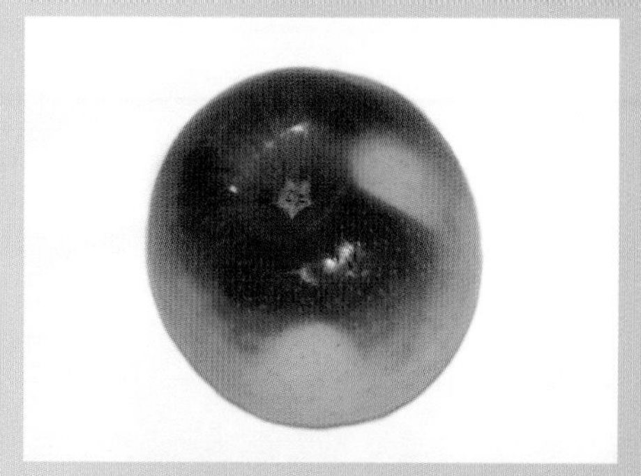

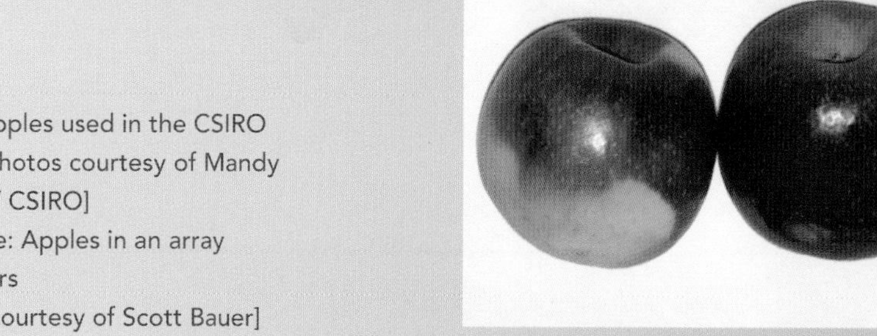

right: Apples used in the CSIRO study [photos courtesy of Mandy Walker / CSIRO]
opposite: Apples in an array of colours
[photo courtesy of Scott Bauer]

Walking on eyes

Sea urchins have no eyes as we know them, but DNA sequencing has revealed that they have genes associated with vision in their tiny, tube-like feet. The discovery was one of several peculiarities revealed by a three-year international project to sequence the sea urchin genome. While sea urchins have been a subject of study for more than 150 years, it wasn't until genetic sequencing was carried out that it was realised the marine creatures had sufficient vision to detect the difference between light and dark.

The animals have proved invaluable to researchers over the decades, particularly in the reproductive field, due to the clearly visible development of the transparent embryo. Observations of how its eggs develop have provided insights into human embryonic growth, including both the formation of identical twins and in vitro fertilisation.

For the sequencing project, involving at least 240 scientists from 11 countries, 814 million pairs of base letters of DNA were decoded from the Californian purple sea urchin, *Strongylocentrotus purpuratus*. The researchers identified 23,300 genes, 7,077 of which are shared by humans.

Among the discoveries that caught the attention of the researchers was the identification of genes associated with a range of human diseases, including muscular dystrophy and Huntington's disease. They hope that by studying the operation of genes in sea urchins, new understanding into human diseases will develop and lead to new treatments, just as study of cell growth in its eggs and embryos led to improved fertility treatment in humans.

The immune system of the sea urchins is another feature that could help in the search for cures for human ailments, especially infectious diseases. Unlike vertebrates with jaws that have an immune system based on antibodies, the sea urchin has a sophisticated "innate immune system" in which proteins detect invading bacteria and warn cells of the presence of unwanted intruders. The complexity of the innate immunity may help explain why sea urchins regularly live to 60 years, and even to 100 years or more.

Genes associated with vision were found in the sea urchins' tube-like feet

opposite: Californian purple sea urchins [photo courtesy of Alex Lin, Eileen Fong, Jin-Hong Kim / California Institute of Technology]

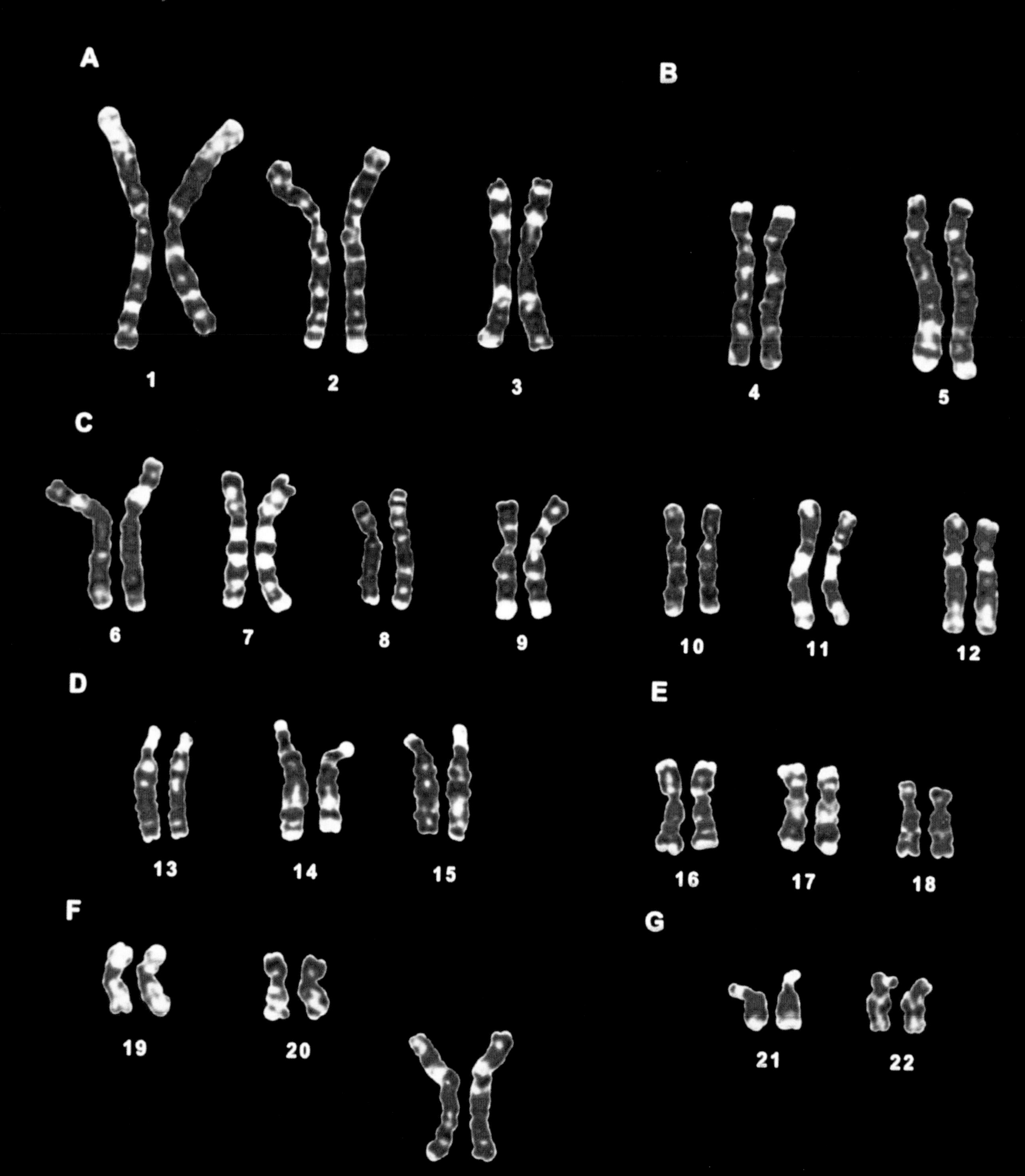
A
B
1
2
3
4
5
C
6
7
8
9
10
11
12
D
E
13
14
15
16
17
18
F
G
19
20
21
22
X X

Book of life

The Human Genome Project launched in 1990 was intended to map, or sequence, all of the genes contained in the human body. When it began, it was estimated that the human genome contained about 100,000 genes. But by the time the last and largest chromosome was sequenced, it was realised that there is a lot less to us than originally presumed. Overall, the project and follow-up work determined there are about 23,000 human genes, many of which are shared by other creatures.

Humans have 23 pairs of chromosomes that are comprised of strands of deoxyribonucleic acid (DNA). Within these are contained the genes, though they only account for about two per cent of DNA, the purpose of the rest remaining largely mysterious. DNA is made up of four chemicals: adenine (A), guanine (G), cytosine (C) and thymine (T). About 3.2 billion pairs of these base chemicals, or letters, are found in the human genome. It is the order, or sequence, of the base letters in the genes that is all-important. The order defines the function, much like the letters of a word define its meaning—anagrams contain the same letters but mean entirely different things.

Decoding the genome was completed in 2003 when the last chromosome, chromosome 1, was sequenced, with the final results published in 2006. In contrast to overall gene numbers, chromosome 1 proved to have more than anticipated with 3,141 genes found, scores more than had been expected. Sequencing chromosome 1 and the other 23 identifies the location of the genes that generate the production, or expression, of the proteins that control basic functions within the body, such as building skin and bones, digesting food and transporting oxygen.

The human genome was mapped as part of a huge international collaboration and with sequencing largely complete, scientists are able to concentrate on identifying the precise role and importance of each of the genes.

After the last and largest chromosome was sequenced, it was realised that there is a lot less to humans than originally thought

opposite: Female human chromosomes
[image: © Science Photo Library]

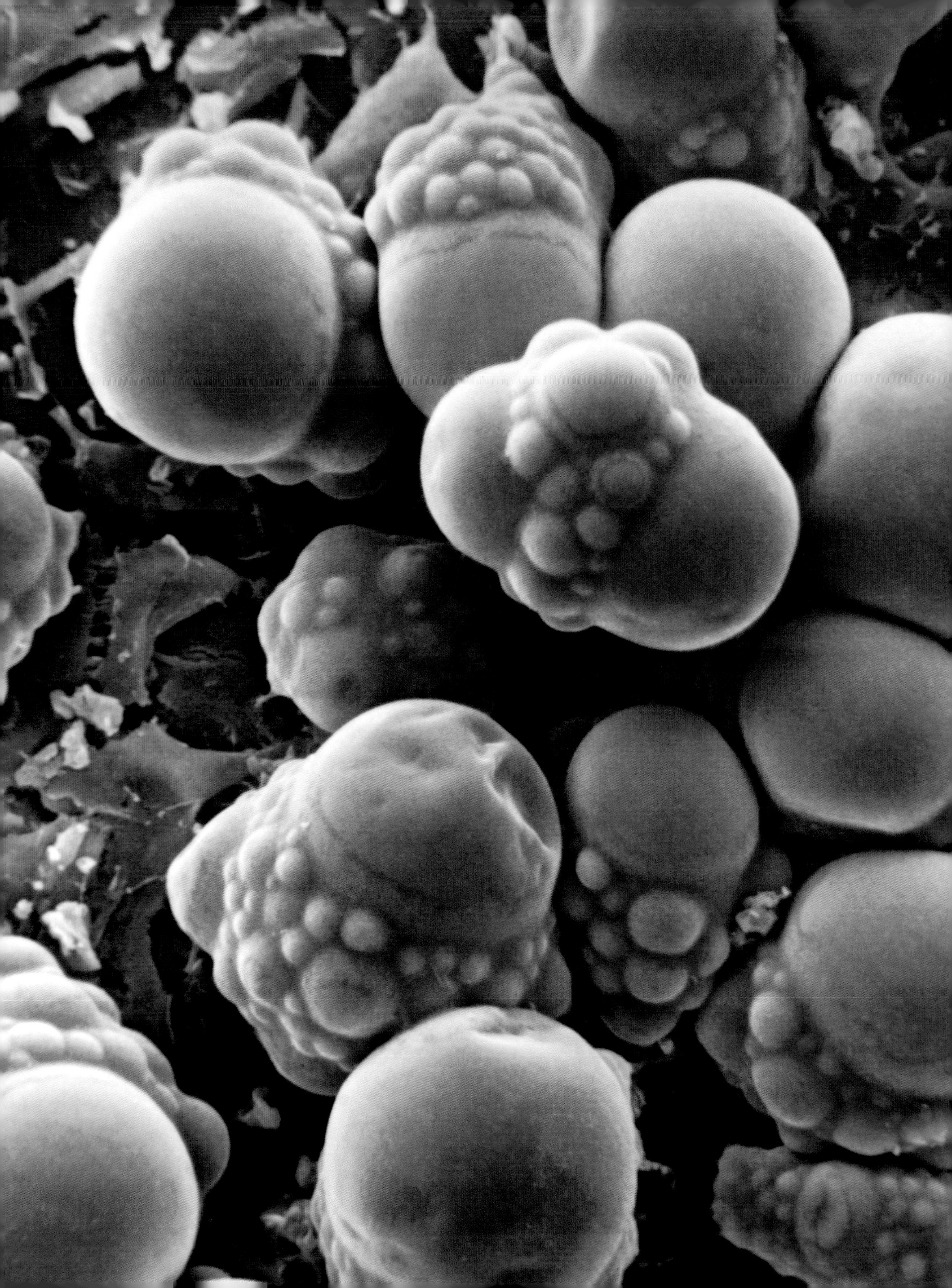

Fat gene

Every failed dieter's perfect excuse was discovered in 2007 when a "fat gene" was identified. Anyone who has certain combinations of the FTO gene has a predisposition to becoming overweight and, therefore, may have to work especially hard to stay slim. People in possession of the worst combination were on average seven pounds (3 kg) heavier than those without it.

There are two variations of the FTO gene, only one of which is associated with an increased risk of being overweight. Because genes are inherited in pairs, with one copy, or allele, coming from each parent, there are four possible pairings. The high-risk group has two alleles that increase the chances of putting on extra weight, the low-risk group has two alleles that confer no extra risk, while the middle-risk group has one allele.

Researchers found that 16 per cent of white Europeans possess two copies of the fat gene, or more accurately the "fat allele", giving them a 70 per cent greater chance of being obese. Similarly, 50 per cent of white Europeans inherit just one copy, which gives them a 30 per cent increased obesity risk. They were found on average to be 2.6 pounds (1.2 kg) heavier than the low-risk group.

The study, carried out by the University of Oxford and the Peninsula Medical School in Exeter, UK, was based on analysis of genetic material from 43,000 people in the UK and Finland. It was picked up during research on diabetes after being spotted as a common denominator among type 2 diabetes patients, and was initially thought to be a factor in the disease. However, after the results were adjusted for weight of patients, it was realised the common link was fat levels rather than diabetes.

Sixteen per cent of white Europeans possess two copies of the so-called fat allele, giving them a 70 per cent greater chance of being obese

opposite: Coloured scanning electron micrograph of fat cells (in pink) from bone marrow tissue [image: © Science Photo Library]

Trawling the genome

The discovery of the so-called fat gene was one of several findings in 2007 from studies funded by the Wellcome Trust in the UK as part of its Case Control Consortium involving about 50 research centres. By surveying the human genome for common factors, rather than identifying an individual group of genes and trying to establish what they do, the researchers identified genetic links to several major diseases.

During one exercise in which 17,000 volunteers provided DNA samples, a team led by Professor Peter Donnelly of the University of Oxford in the UK was able to identify an unprecedented number of genes linked to common diseases. It was the biggest human genome survey to date and pinpointed 24 genetic variants, which are carried by up to 40 per cent of the UK population and influence six diseases.

Genes associated with heart disease, rheumatoid arthritis, type 1 and type 2 diabetes, Crohn's disease, and bipolar disorder were identified, with high blood pressure a possible seventh. One of the most valuable aspects of the genome association technique is that it can locate common genes that carry a small level of increased risk, rather than, as has been done previously, find rare genes that give carriers a high risk of developing a disease.

The biggest human genome survey to date pinpointed 24 genetic varients that influence six diseases

below: Computer screen showing a sequence of base pairs forming part of the human genetic code [image: © David Parker / Science Photo Library]

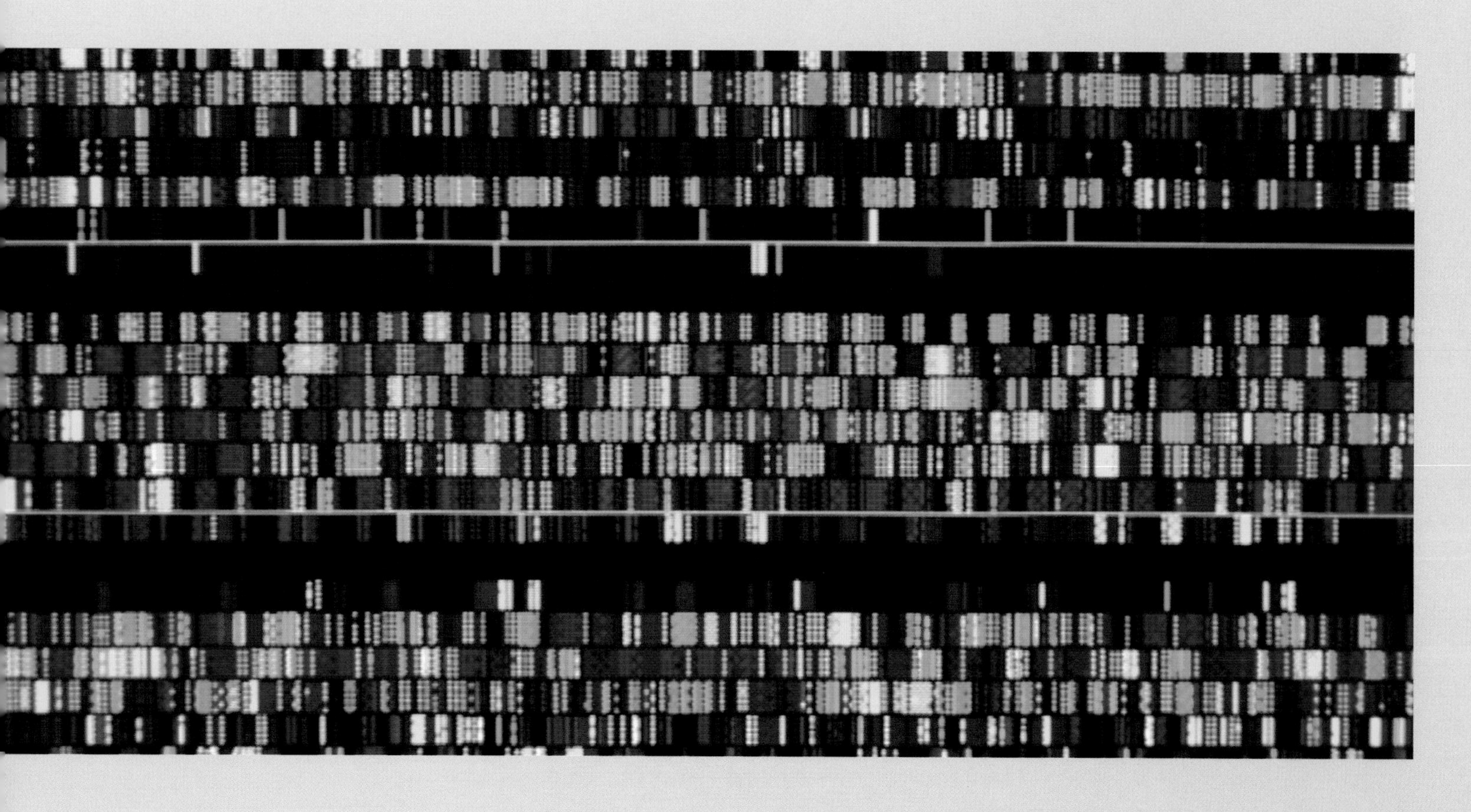

Common cancer

Genes associated with an increased risk of developing breast cancer were found in a similar but separate study led by Professor Bruce Ponder of the University of Cambridge in the UK. By scanning large sections of the genome—in this case 200,000 blocks of DNA—his team was able to pinpoint four genes linked to the cancer and identified the location of a fifth. In one piece of research the scientists managed to identify almost as many genes linked to the onset of breast cancer as had been found in the previous 14 years. It was found that women who had a mutated form of one of the genes had a one in 10 chance of developing breast cancer by the age of 70, compared to the average risk of one in 18.

One of the most significant findings of the study was that the genes identified are common in the female population, thus having implications for a wide sector of society rather than just a handful of people. Earlier research had identified other genes, notably the high-risk BSCA1 and BRCA2, which, while condemning about 80 per cent of the women with mutated variants to develop breast cancer, are only found in about one in 500 women. All the newly identified genes are spread widely through the population and are thus likely to have an accumulative effect. Individually they will only cause a small increase in risk, but many women are likely to have more than one of the genes.

The breast cancer gene study, along with those coordinated by the Wellcome Trust and other organisations, raise the prospect of risk assessment screening for individual patients. Once all the risk factors of various genes have been established for a disease, there would be the potential to test a patient for all the genes associated with increased risk. Such a health check system could assess people for the risks of developing a range of diseases.

All the newly identified genes are spread widely through the population and are likely to have an accumulative effect

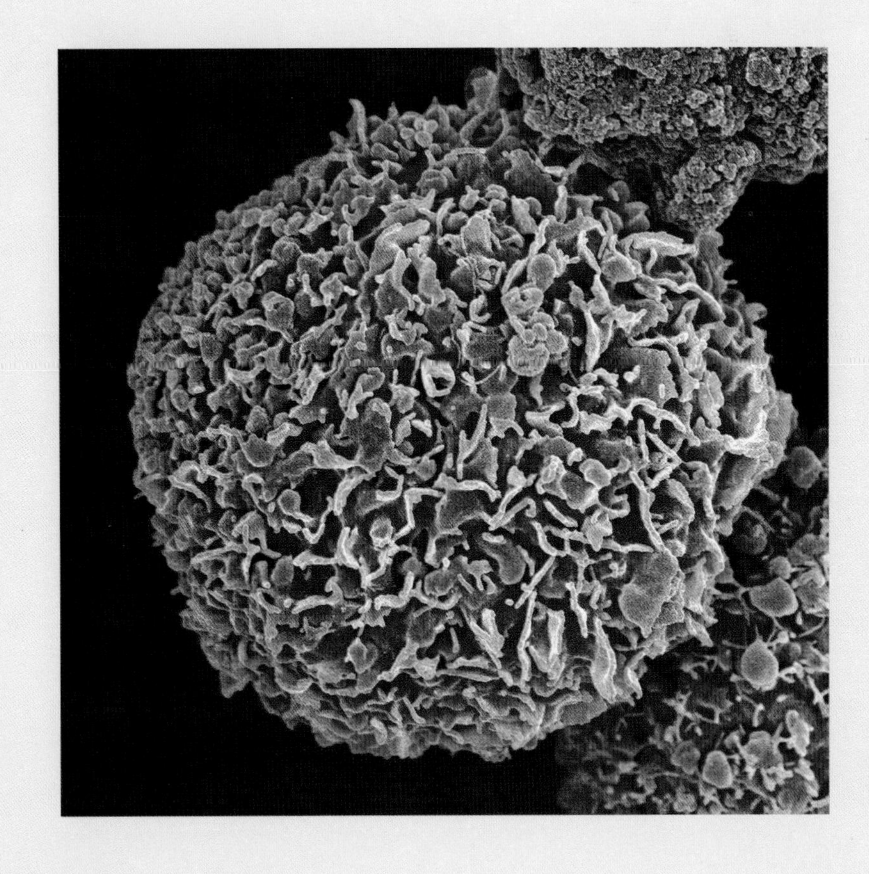

right: Coloured scanning electron micrograph of a breast cancer cell [image: © Steve Gschmeissner / Science Photo Library]

Junk between the genes

Mutations in junk DNA may have just as many consequences as those in genes

The large quantity of genetic material in between genes has been described as junk DNA, but is likely to play a much more important role than hitherto expected, according to a US-led study. Researchers attempting to put together the first "parts list" to describe all the functional elements of one per cent of the genome have found that most junk DNA is, after all, biologically active.

The four-year study challenges the traditional view of genes as the only elements of the genome that really matter, suggesting that there are many types of sequences that play a role. Close analysis of one per cent of the genome showed that the junk DNA helped the genes produce proteins.

The study found that the majority of junk DNA is encoded into the ribonucleic acid (RNA), ensuring the signals sent by genes for the formation of proteins are taken to the right places and carried out.

The analysis was led by the European Molecular Biology Laboratory's European Bioinformatics Institute (EMBL-EBI), based in the UK, as part of work for the ENCyclopedia Of DNA Elements (ENCODE), an international research consortium led by the National Human Genome Research Institute (NHGRI) in the US. The findings have implications for the search for cures and treatments for diseases because mutations in the junk DNA might have just as many consequences as mutations in genes.

below: Computer model of a double-stranded RNA molecule [image: © Laguna Design / Science Photo Library]

Mapping a

It has already proved itself resistant to several antibiotics and there are fears it will develop defences against others

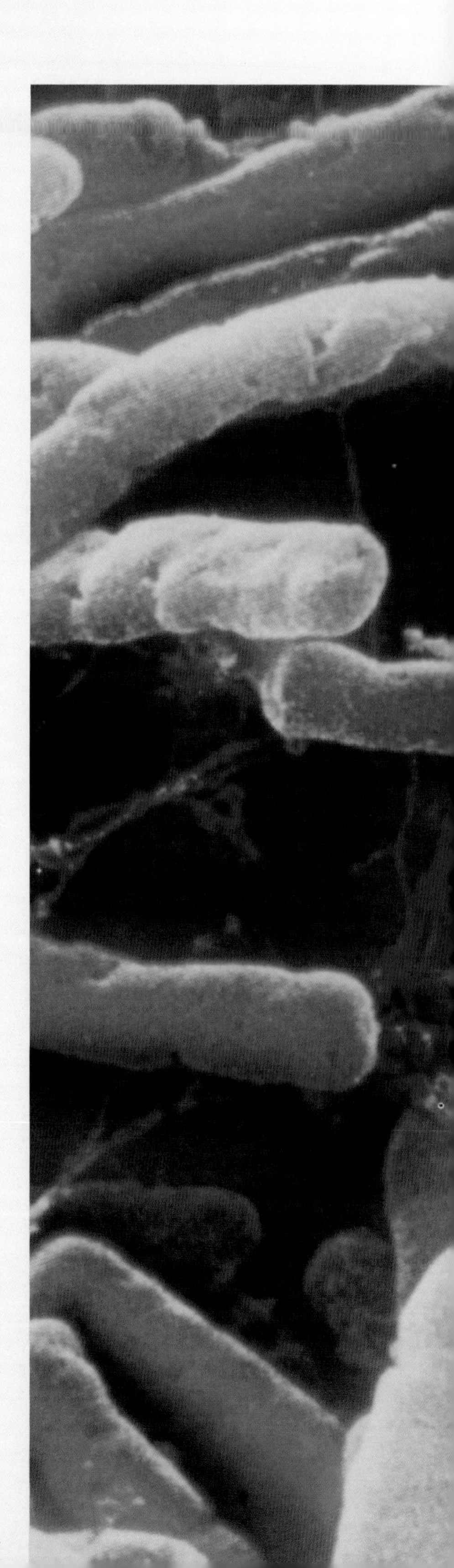

In an effort to seek out the weak points of the hospital superbug *Clostridium difficile*, researchers have mapped its genome. *C. difficile* is the major cause of infections acquired by patients while in hospital and is more deadly than MRSA, another notorious superbug. It is a species of bacteria that has already proved itself resistant to several antibiotics and there are fears it will develop defences against others.

Researchers from the Wellcome Trust Sanger Institute in the UK mapped the bacterium's genetic code in the hope that by shedding light on its structure, a more effective means of combating it can be developed. They found that more than 10 per cent of the genome is made up of sequences that can move from one organism to another. In this way, it can acquire genes that improve its resistance to antibiotics and help it thrive in the human gut.

Its closest relatives include the bacteria that cause botulism, tetanus and gas gangrene, but the research showed that it shares only half its genes with them. Moreover, it has wide genetic variability between its own strains. Among the findings yielded by the mapping project was the discovery that *C. difficile* produces the chemical paracresol to kill other bacteria to make more room for itself in the human body.

right: Coloured scanning electron micrograph of *C. difficile* bacteria [image: © D. Phillips / Science Photo Library]

superbug

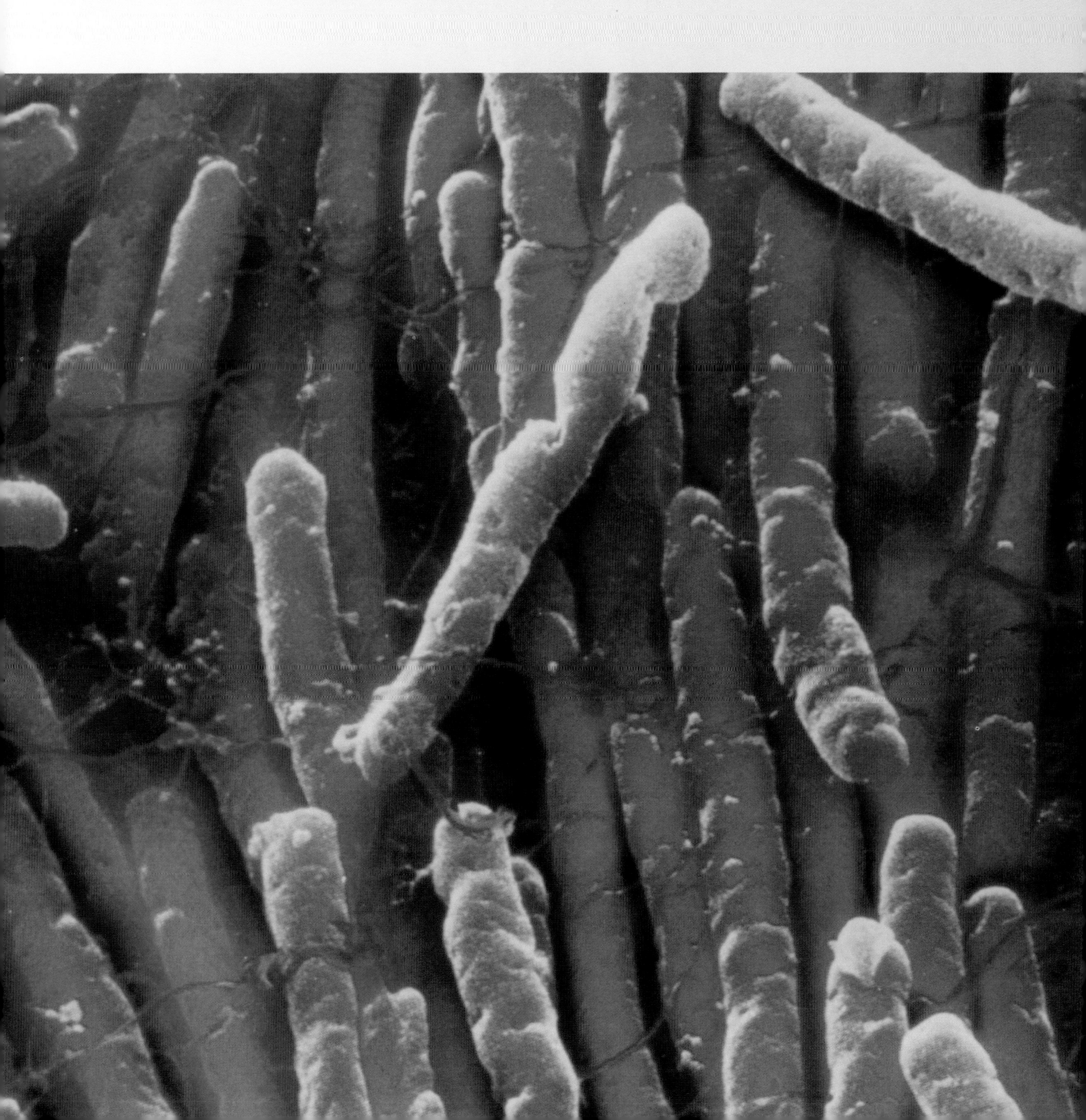

The white stuff

above: Neolithic skull used in the study at Mainz University [image courtesy of Joachim Burger, Palaeogenetics Group]

A study of Neolithic bones has shown that the ability among adult North Europeans to drink milk evolved within the last 8,000 years. To the majority of people around the world, a glass of milk is a fast route to diarrhoea, stomach cramps and bloating—but among North Europeans it has over time been a survival aid. The key to being able to digest milk comfortably as an adult was the development of a gene mutation conferring lactose tolerance. Most people and other mammals lose the ability after being weaned. It works by producing the lactase enzyme that breaks down lactose, one of the prime sugars contained in milk.

Researchers at Mainz University in Germany and University College London in the UK have determined that the lactase gene mutation only appeared within the last few thousand years and then spread rapidly. More than 90 per cent of people of North European origin have the gene, as do some African and Middle Eastern populations. The speed with which it spread through most of the North European population suggests that natural selection must have been a strong influence, and that possession of the gene was a significant advantage to survival. Being able to drink milk would have meant, the researchers argued, that people who had access to dairy livestock would have been able to consume a fortifying drink virtually on demand, cushioning them from the seasonality of other foods. An added advantage is that it would have been free of parasites contaminating water supplies.

Scientists reached their conclusions after finding that the gene was missing from Neolithic skeletons dated at 5840 to 5000 BC, from some of the earliest organised farming communities in Central, Northeast and Southeast Europe. The finding, they said, supports the theory that lactose tolerance came about because of exposure to milk after dairy farming began in Europe about 9,000 years ago, and challenges the alternative idea that the ability to drink milk led people to keep dairy animals.

The key to being able to digest milk comfortably as an adult was the development of a gene mutation confering lactose tolerance

Burning brighter longer

A study of people with an average age of 99 revealed that those with the CETP VV gene are twice as likely to still be mentally alert as those without it

A gene already associated with long life has been found to ensure the retention of a sharp, incisive mind in old age. While others around them are losing their marbles, elderly people who inherit the CETP VV gene are far more likely to remain clear-thinking even beyond their 100th birthdays. A study of 158 people aged at least 95, and with an average age of 99, revealed that those with the gene are twice as likely to still be mentally alert as those without it. In tests, 61 per cent of those with the gene were judged to be mentally alert, compared to 30 per cent of their peers without the gene.

Furthermore, the gene variant appears to provide some measure of protection against Alzheimer's disease and other types of dementia. When researchers from Yeshiva University in the US assessed a further group of 124 people aged 75 to 85, it was found that they were five times less likely to suffer dementia than elderly people with a different variant of CETP. Dr Nir Barzilai led the study for the university's Albert Einstein College of Medicine, having previously established that the CETP VV gene variant is almost three times as common among centagenarians as it is in people still in their 60s.

It is thought that the CETP VV variant helps ensure long life by producing larger cholesterol particles than other CETP genes. Because the cholesterol is bigger, it is believed to be less likely to get lodged in blood vessel linings and thus avoids the clogging that leads to heart attacks and strokes. It may perform a similar anticlogging function in the brain, helping to ward off dementia.

left: Dementia patient and granddaughter [image: © Henny Allis / Science Photo Library]

Sugar and spice

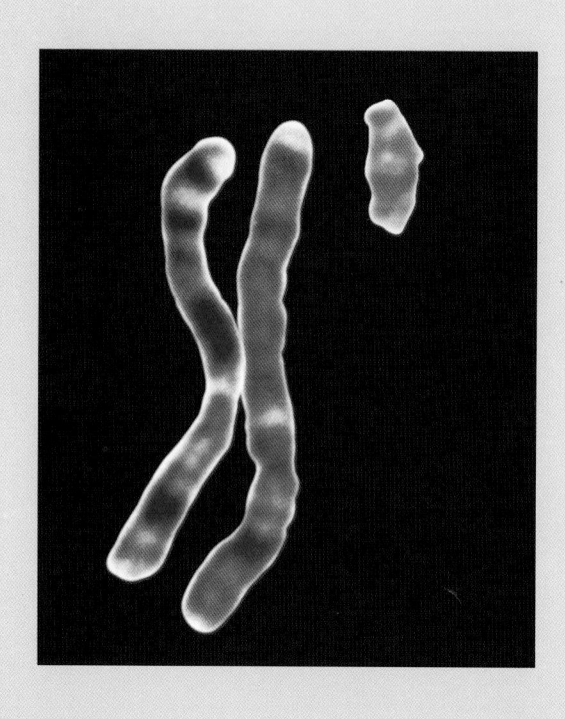

Four brothers from southern Italy discovered that they should have been born as girls

Genetic studies have shown that gender is more complicated than simply juggling X and Y chromosomes. Researchers investigating four brothers from southern Italy have discovered that they should have been born as girls, except that a crucial gene was missing. Instead of having one X and one Y chromosome, as is usual for males, the brothers were each found to have two X chromosomes—the combination expected for females.

In 1990 the SRY gene was discovered to be a primary trigger of male development, but it was absent from the brothers. Instead, the team led by researchers from the University of Pavia in Italy identified the gene that prompts an embryo to develop into a girl. In the case of the four brothers, all of whom are infertile, the two copies of the RSPO1 gene inherited by each of them had mutated. The researchers concluded that in order for a girl to develop in the womb at least one of the RSPO1 genes must be fully functional.

above: Magnified image of the X and Y chromosomes of a human male
[image: © Department of Clinical Cytogenetics, Addenbrookes Hospital / Science Photo Library]

Family history

Analysis of the Y chromosome has enabled researchers to find ancestral links between complete strangers. Among the most memorable links discovered was between Genghis Khan, the Mongol warlord whose name is still associated with killing and destruction centuries after his death, and an accountant in the US. Tom Robinson, at the time an associate professor in accountancy at Miami University, was shown through checks on his Y chromosome to be descended from an ancestor of the thirteenth-century warlord.

Similar techniques have been used to establish genetic links between living people who share the same surname but have no known family connections. Researchers at the University of Leicester in the UK brought together a group of 300 male volunteers who shared 150 surnames. Each same-name pairing was assessed for similarities in the Y chromosome, which is passed from father to son almost unchanged even over many generations. It was found that almost a quarter of the pairings shared an ancestor who lived within the last 20 generations. When common names such as Smith and Jones were excluded from the calculations, the chances of a pair being related rose to more than a third.

Professor Mark Jobling, who led the research, said the technique had the potential to help police solve crimes by providing a list of the likely surnames of suspects. A database in the UK of 40,000 surnames and the Y chromosomes associated with them would be needed, he said, and would offer a name match in one in five crimes where DNA was found at the scene.

A link was discovered between Mongol warlord Genghis Khan and an American accountant

right: Portrait of Genghis Khan, founder of the Mongol Empire (1260-1368) [image: Bridgeman Art Library / Getty]

Not all that we seem

John Revis had thought his line of descent was as English as roast beef and Yorkshire pudding, but his Y chromosome hinted at a more exotic ancestry. When he provided a DNA sample after a plea for volunteers by geneticists at the University of Leicester in the UK, his Y chromosome proved to be a rare variant previously only seen among men of West African origin.

Researchers, again led by Professor Jobling, tracked down and tested the DNA of a further 18 men with the name Revis. Six of them shared the rare hgA1 variant and all were white. Until the Revis samples were taken only 26 people worldwide had been found to share the hgA1 Y chromosome. Of these, 23 were from West African countries including Senegal and Guinea-Bissau, and three were African Americans.

The researchers attempted to trace the family trees of seven Revis samples, including that of one man in the US but whose ancestor emigrated from the UK in the nineteenth century. The paper trail ended in about 1780, when the samples had been traced to two family trees in Yorkshire, and it is suspected that they converged in the early eighteenth century.

When the distinctive chromosome first arrived in the UK remains unclear. Professor Jobling said it was most likely a legacy of the slave trade—by the late eighteenth century there were 10,000 black people living in the UK. It is possible, however, that it arrived with the Roman occupation of Britain, with historical records showing that a garrison of soldiers from North Africa reached British shores in about 200 AD.

John Revis thought his lineage was as English as roast beef, but his Y chromosome told another story

Wild at heart

above: Domestic cat [photo courtesy of Josette Gerlier]

The domestic cat would doubtless have us believe that for time immemorial its natural place has been in the home with a human on hand to supply it with food, warmth and—when it's in the mood—affection. Archaeological finds suggest that cats have lived with people for at least 9,500 years and it has long been assumed they were originally wildcats. Researchers have now pinpointed the Near Eastern wildcat, rather than the European wildcat, as the animal that originally recognised the advantages of living alongside humans. It is thought that the wildcat that first came to live with people did so in the Fertile Crescent associated with the Nile, Tigris, Jordan and Euphrates rivers in North Africa and the Middle East.

Research found that all domestic cats are descended from five female Near Eastern wildcats, *Felis sylvestris lybica*, and that they go back 131,000 years when the subspecies became distinct from the European wildcat, *F.s. silvestris*. Professor David Macdonald, who led the study, said it was likely that the wildcat domesticated itself, lured to human settlements by the rodents that infested grain and other food stores as people turned from a hunter-gathering to an agricultural lifestyle.

To carry out the study, genetic samples were taken from live domestic cats and wildcats from Europe, Africa, the Middle East and Asia. Further wildcat samples were provided by museums. By analysing the mitochondrial DNA, which in mammals is inherited only from mothers, the team was able to build up a clearer idea of the cat family tree. They established that the African wildcat justified being regarded as two distinct subspecies—the Southern African wildcat, *F.s. cafra*, which inhabits central and southern regions of the continent, and the Near Eastern wildcat.

More controversial was the finding that the Chinese desert cat, *F.s. bieti*, should be included within the wildcat grouping. It separated from the European wildcat 230,000 years ago. Members of the seven-nation research team accepted that, while the genes showed it to be a wildcat, there will be disagreement among other scientists because of clear differences in the animal's body shape. Samples from the sand cat, *F. margarita*, confirmed that the creature is a different species from the wildcat, though they share a common ancestor.

Professor Macdonald, the director of the Wildlife Conservation Research Unit at Oxford University in the UK, was confident the DNA study would help conservation work with wildcats. He is particularly concerned about the European wildcat in the UK, where it is classified as the Scottish wildcat. Estimates put the population at 300 to 400, but it has been impossible for conservationists to be sure due to the level of interbreeding with domestic cats. Now that the genetic markers for the subspecies have been identified, it should be possible to provide a more accurate picture of the numbers and geographic location of European wildcats still surviving in the UK and other areas. It should also help reveal just how acute is the problem of interbreeding with the domestic cat, an animal that has spread across the world so successfully it was described by Professor Macdonald as one of the most prolific biological experiments people have ever carried out.

All domestic cats are descended from five female Near Eastern wildcats

Elastic dogs

A genetic variation that helps explain why man's best friend can be as big as a pony or as small as a kitten has been isolated in dogs. Dogs have among the most elastic body size of all creatures, with the biggest being 70 times heavier than the smallest. Researchers have identified a section of DNA that they believe is a key element in the processes that enabled the wolf to turn into both a Chihuahua and a Great Dane. Wolves are thought to have been first domesticated by man 12,000 to 15,000 years ago, and they have since descended into dogs of many shapes and sizes.

In research conducted by scientists from the US and the UK, it was discovered that all dogs under 20 pounds (9.1 kg) have a sequence of DNA that reduces the influence of a growth gene. The mutated sequence lies next to the growth gene IGF1 and inhibits the production of a protein hormone that prompts growth. By restricting the impact of the growth gene, the DNA sequence keeps dogs small. The IGF1 gene is found in medium and large dogs, but the mutated sequence is absent in all except for Rottweilers. Other genes and gene regulators are expected to explain why the Rottweiler breaks the rule.

Portuguese water dogs formed the basis of the study and once the DNA sequence variation, or haplotype, was identified, the research was expanded to assess the genetic composition of 3,241 dogs from 143 breeds. Among the breeds analysed in the study were some of the very smallest dogs produced from selective breeding, such as toy poodles, Chihuahuas, pugs and Pekinese, as well as some of the biggest, including Irish wolfhounds, mastiffs, Great Danes and Saint Bernards. Professor Gordon Lark, of the University of Utah, was one of the lead researchers and said the Portuguese water dog was an ideal starting point because it ranged in size from 25 to 75 pounds. His interest was further encouraged because he was a water dog owner.

The origin of the genetic mutation found in small dogs is thought either to have been present in one or more unusually small wolves that were among those that were domesticated, or to have occurred at the beginning of the domestication process. It is absent in today's wolves. Dr Elaine Ostrander, director of the National Human Genome Research Institute, said learning what dictates size in dogs provides insights into what determines human height and the growth of cells.

Dogs have among the most elastic body sizes of all creatures, with the biggest being 70 times heavier than the smallest

opposite: An Irish wolfhound and a Chihuahua mix [photo courtesy of Tyrone Spady]
right: An Afghan hound and a Chihuahua mix [photo courtesy of Edouard Cadieu]

previous, clockwise from top left: Artist's concept of Cassini [image courtesy of NASA / JPL / University of Arizona], the Mars Express orbiter [illustration courtesy of NASA / JPL-Caltech], artist's concept of the twin Mars Exploration Rovers, Spirit and Opportunity [image courtesy of NASA / JPL-Caltech], Enceladus [image courtesy of NASA / JPL-Caltech], artist's concept of plant life on a planet that orbits a different class of star [image courtesy of Doug Cummings, CalTech], artist's impression of Gliese 581c [illustration courtesy of ESO], the Crab Pulsar [image courtesy of NASA/CXC/ASU/ J. Hester et al., HST / ASU / J. Hester et al]

Stars, Planets & Space

Just gazing up at the night sky is enough to evoke a sense of wonder about the stars, galaxies and the Universe. Built on such a vast scale and over such awesome distance and time, they seem barely comprehensible. But since the first days of civilisation people have been trying to make sense of space and mankind's place in it.

Apart from the straightforward desire to understand our surroundings, the study of space is a search for insights into our own creation. Astronomers no longer view the Earth or even its solar system as being at the centre of the Universe—but it is still the only place among the stars where life is known to exist. Life itself and how and why it started is one of the mysteries of the Universe; finding out where else it may be found, or once existed, has long been an aim of scientists.

By uncovering the laws of physics that govern the Universe, the history of the stars and the galaxies and the conditions in which they and the planets were formed, as well as the conditions necessary for life to start, scientists hope to eventually explain why we are here and whether we are alone.

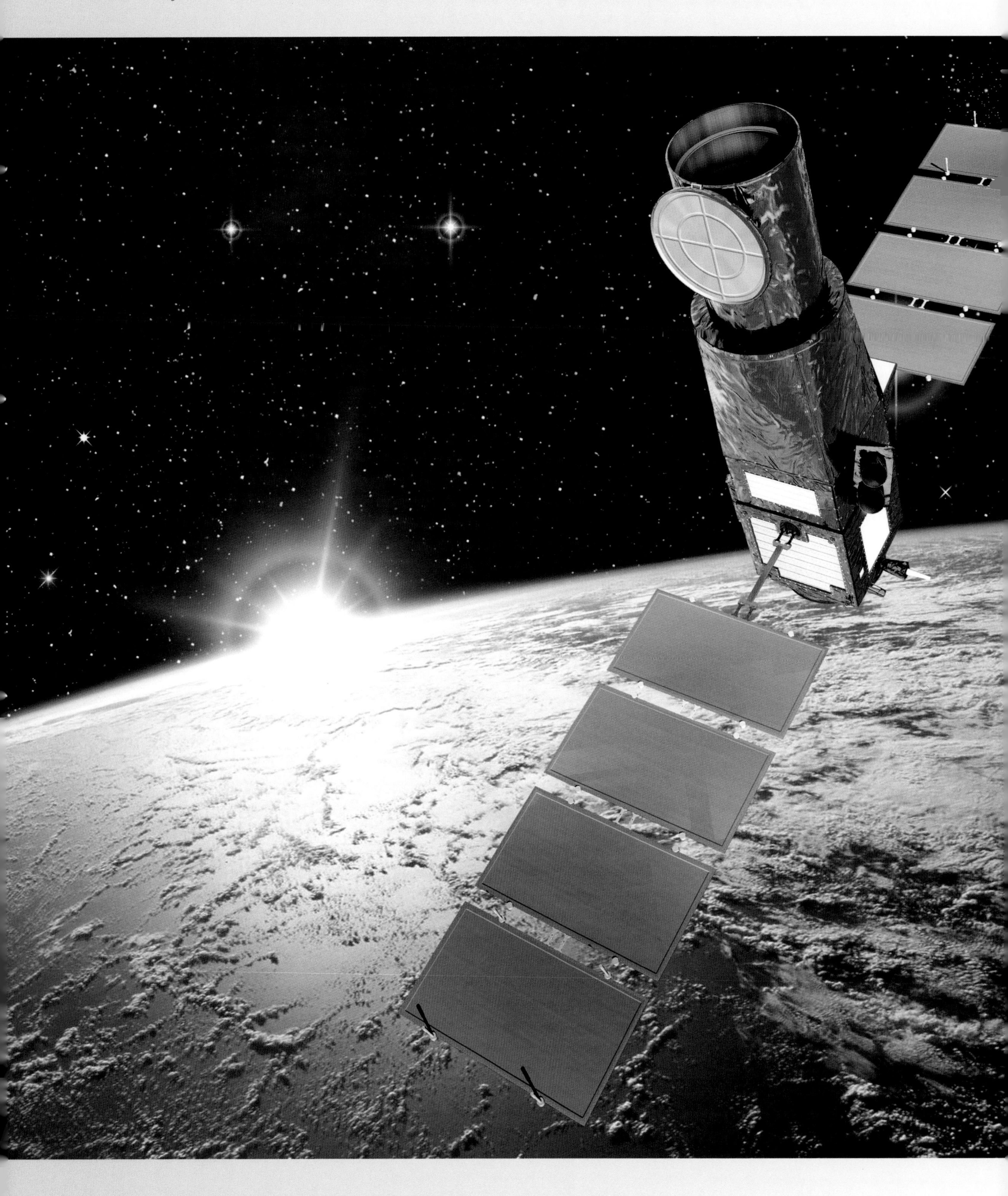

Planet hunting

While it wasn't until the 1990s that a planet outside our own solar system was discovered, there have since been regular announcements of new finds, with well over 200 extrasolar planets recently identified.

All of them are larger than Earth—the vast majority being many times bigger—and many are gas giants even more massive than Jupiter. The search for rocky, Earth-like planets that could harbour life, and perhaps one day provide an alternative home for mankind, is indeed well underway.

Observatories continue to pinpoint new planets and at the end of 2006 the COROT (Convection Rotation and planetary Transits) space probe was launched to orbit Earth in an attempt to find many more. COROT, a French-led mission, is designed to measure the light given off by stars and detect slight dips in brightness, which indicate the presence of a planet in transit across its face.

The first that it discovered, in May 2007, was of a planet 1,500 light years away in the Monoceros (Unicorn) constellation. A hot gas giant 1.3 times bigger than Jupiter, it was named COROT-Exo-1b and takes just 1.5 days to orbit its parent star.

opposite: Artist's depictions of the COROT space probe [illustrations © Sciencec Photo Library / D. Ducros]
above right: Artist's depictions of the COROT space probe [illustrations courtesy of CNES / D. Ducros, © 2006]

The search for Earth-like planets that could harbour life is well underway

Goldilocks

planet

Not too hot, not too cold, but just right for life

The first small extrasolar planet to be found in a star's "Goldilocks zone"—not too hot, not too cold, but just right for life—was announced little more than a week earlier than COROT's first find.

Gliese 581c was pinpointed by observations from the European Southern Observatory at La Silla in the Atacama Desert in Chile by a team of Swiss, French and Portuguese scientists. Dr Stéphane Udry of the Geneva Observatory in Switzerland was one of the lead researchers and said the discovery of Gliese 581c, which orbits its star in 13 days, was a step towards finding a planet suitable for humans.

The planet is bigger than Earth, being 50 per cent greater in diameter, but is considered more likely to harbour life than any planet yet discovered. Surface temperatures on Gliese 581c were calculated to be somewhere from 0 to 40°C (32 to 104°F), meaning that any water—regarded as a prerequisite for life—would be in liquid form on the surface. The planet's parent star, Gliese 581, is a red dwarf that is considerably cooler than our own sun, but the planet is 14 times closer to its star than Earth is to the Sun. Another factor making the planet a good candidate in the search for life outside Earth is the age of the star, which, at several billion years old, has given life plenty of time to evolve. The prospects are increased by the red dwarf's comparatively low level of activity, meaning the planet avoids the high radiation levels that would kill off emerging life.

At 20.5 light years away, getting there is at present a distant dream if not an outright impossibility. To travel there in the fastest spacecraft yet launched, the New Horizon craft on its way to Pluto, it would take more than 250,000 years to reach Gliese 581c, or more than 750,000 years if using the slower Space Shuttle. But finding it doubled the number of small planets—now Earth and Gliese 581c—known to be at habitable distances from stars, and thus indicated that many more are likely to be awaiting discovery.

opposite: Artist's impression of the planetary system around Gliese 581, with planet Gliese 581c in the foreground
below: The star Gliese 581 [all images courtesy of ESO]

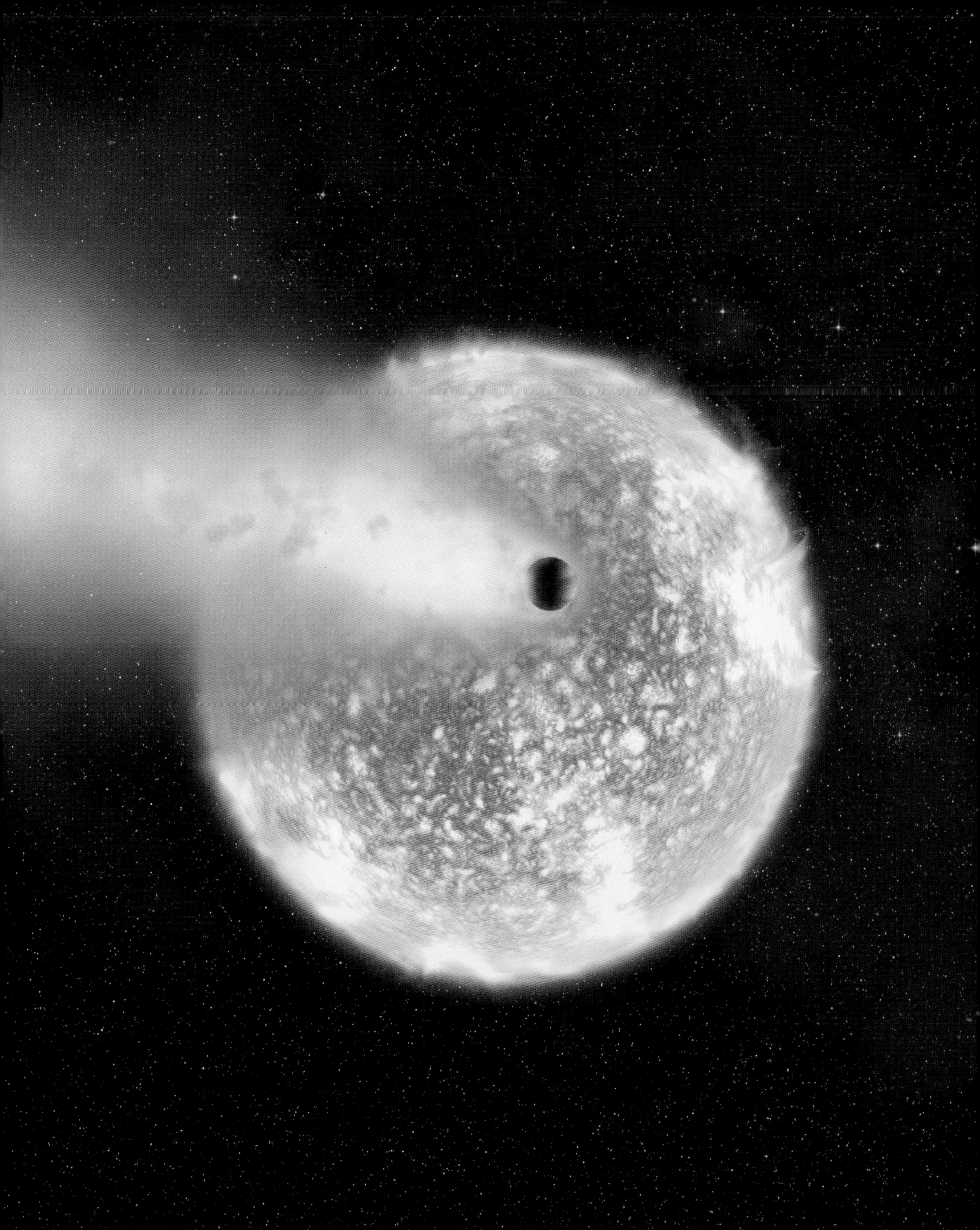

Reading the atmosphere

The first extrasolar planet identified as having water content is a hot gas giant with surface temperatures of 700°C (1,292°F). Osiris, officially called HD 209458b, lies 150 million light years from Earth and has been the subject of a number of studies to detect the contents of its atmosphere. Researchers at the Lowell Observatory in the US determined that the atmosphere contained water by analysing theoretical models and observations from NASA's Hubble Space Telescope.

The detection of water is an important guide to the likelihood of life being found on extrasolar planets, as it is assumed that there will have to be a supply of water for any chance of life to thrive. The composition of Osiris's atmosphere can be detected because it is one of a small proportion of extrasolar planets that, from the Earth's line of vision, cross the face of their stars. By detecting differences in the light spectrum during and after such a transit, scientists are able to calculate the composition.

Jeremy Richardson of the NASA Goddard Space Flight Centre carried out one of the studies of Osiris using a spectrograph on the Spitzer Space Telescope. While there was no expectation of detecting suitable conditions for life on Osiris, he said such techniques could prove invaluable in trying to assess the chemical composition of extrasolar planets and in providing clues to how they were formed.

The detection of water is an important guide to the likelihood of life being found on extrasolar planets

opposite: Artist's depiction of the atmosphere of Osiris being "boiled" away by its parent star, HC 209458 [image courtesy of ESA / Hubble]
right: Artist's concept of a cloudy hot gas giant [image courtesy of NASA / JPL-Caltech]

An extra pole

Observations of the Crab Nebula have thrown up the entirely unexpected discovery of a pulsar that appears to have three magnetic poles. The finding challenges the previously held assumption that cosmic bodies, whether the Earth or distant stars, have just two poles.

The intriguing anomaly was found at the centre of the Crab Nebula, the remains of an exploded star 6,300 light years from Earth. The explosion was seen and recorded on Earth by Arabic and Chinese astronomers in 1054, when it was so bright that it could be seen during daylight. At the centre of the nebula's supernova—which is expanding at more than 930 miles (1,497 km) per second—is a pulsar formed of the supercompressed remnants of a giant star. Unexpected blasts of energy from the pulsar, or neutron star, were recorded by the Arecibo Observatory in Puerto Rico and suggested the existence of a third pole. If this is the case, it would make the pulsar the first cosmic body to have three magnetic poles.

The pulsar, which rotates 30 times a second, emits radiation in two narrow beams that are thought to have been focused by the magnetic poles. Such emissions are predictable but, to the consternation of astronomers, extra blasts of energy have been detected close to the anticipated pulses. Professor Tim Hankins, of the observatory, and Professor Jean Eilek, of the New Mexico Institute of Mining and Technology in the US, discovered the unusual pulses. They suggested the energy emissions could be caused by a third magnetic pole and said that it raised a host of questions about pulsar radiation patterns.

The pulsar at the centre of the Crab Nebula was one of the first to be discovered and its radiation has been used in mapping the Sun's corona and measuring the atmosphere of Titan, Saturn's largest moon. Radio signals from the pulsar can last as little as four-tenths of a nanosecond, yet can account for as much as a tenth of the energy emitted by the Sun.

Radio signals from the pulsar can last as little as four-tenths of a nanosecond, yet can account for as much as a tenth of the energy emitted by the Sun

opposite: Hubble Space Telescope image of the Crab Nebula [image courtesy of NASA, ESA, Allison Loll & Jeff Hester / Arizona State University & Davide de Martin, ESA / Hubble]

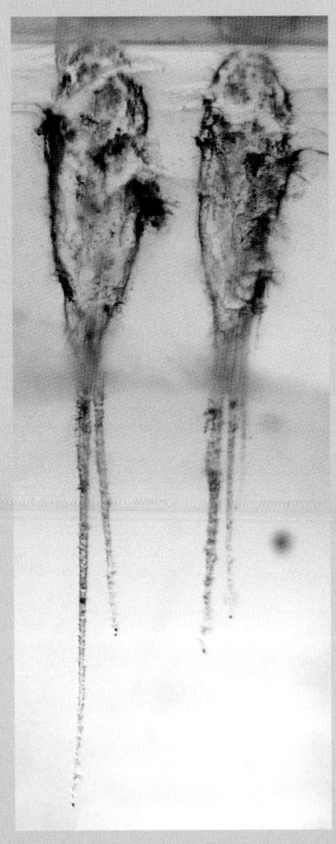

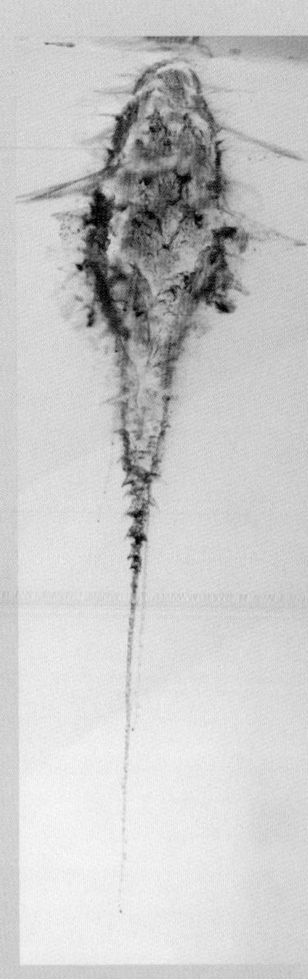

left: Comet particle tracks caught in the stardust aerogel [image courtesy: NASA]
below: Artist's concept of comet Wild 2 as seen from NASA's Stardust spacecraft during its flyby on 2 January 2004
[image courtesy: NASA / JPL-Caltech]

Stardust

Specks of dust dating back 4.57 billion years have been yielding insights into the creation of the solar system. A thousand grains of dust were collected during a NASA mission to the Wild 2 comet, which is thought to have been created within the first 10 million years of the birth of the solar system.

Materials found within the comet's dust included the water elements and organic compounds that were essential for life to begin on Earth

By capturing specks of material that were around during the birth of the Sun and the planets, scientists hoped to get a better idea of the processes involved in the creation of a solar system. Among the questions researchers hope to answer by studying the stardust is how exactly planets are formed from clouds of dust and gas. One speck of dust, so tiny that its precise content could not immediately be determined, was found to predate the solar system. It is rich in oxygen and would have formed an infinitesimal part of the cloud of dust and gas that formed the building blocks for the solar system.

NASA's Stardust probe was launched in February 1999, collected samples from the comet in January 2004, and landed back on Earth in January 2006 with its unprecedented cargo. Scientists from around the world have since been involved in researching the composition of the dust collected from Wild 2.

Within just a few months after research commenced, the dust was already throwing up surprises. Among them was the realisation that the solar system in its infancy must have been far more volatile than previously thought. The dust shows that the system's building blocks were far more intermingled than first assumed, which raises new questions about how the planets formed and why they have different compositions. Until the Stardust payload was analysed, the main theory as to why planets differed in their composition was that they were in different parts of the system, which contained distinct building materials. But some of the compounds found in the dust could only have formed in places where there was considerable heat, which showed that a tenth of the comet must have formed much closer to the Sun than had been suspected.

Other materials found within the dust included the water elements and organic compounds that were essential for life to begin on Earth. Among them were two types of nitrogen-rich organic molecules that had never been seen in comets before, and a new class of organic substances. Wild 2 was preserved as a relic of the birth of the Sun and planets because it spent most of the last 4.57 billion years in the outer reaches of the solar system, where it was kept in isolation in deep-freeze conditions. It was kept in pristine condition, safe from changes caused by solar heat or space debris from the planets, until it was eventually hauled out of its orbit by Jupiter's gravitational pull. Dust samples were collected in 132 ice cube-sized blocks of aerogel—a silicon-based porous solid dubbed "frozen smoke", which has the lowest density yet created for a solid. Aerogel's low density meant that as the probe flew through the comet's tail at 13,000 miles per hour (21,000 km/h), the dust was slowed, halted and trapped without being vaporised.

Live coverage of a star's death

The explosion was so bright it outshone its entire galaxy

Astronomers were given an unprecedented view of a star's demise when a NASA spacecraft picked up a telltale burst of gamma rays. It was the first time a supernova was witnessed from such an early stage of the star's death throes, and the explosion was so bright it outshone its entire galaxy. They were alerted to the extraordinarily powerful explosion when the Swift spacecraft detected an unusually long gamma-ray burst (GRB) that was ten million billion times as powerful as the Sun.

Supernovas have been detected in the past, but never before at such an early stage and in such detail as on February 18, 2006. It is believed the gamma-ray burst, in the form of a jet of high-energy X-rays, was emitted as the core of the star collapsed. The rest of the star imploded a few minutes later to create the massive supernova explosion, including a bubble of gas produced by the shock wave and heated to two million degrees Celsius (3.6 million degrees Fahrenheit).

Gamma-ray bursts, which Swift was designed to detect, usually last from a fraction of a second to a few tens of seconds—but on this occasion the burst lasted 40 minutes. Swift directed all three of its instruments to observe the event and alerted ground-based observatories to join in within 20 seconds. The star was 440 million light years away and the Very Large Telescope at the European Southern Observatory in Chile confirmed two days later that an exploding star caused the gamma-ray burst.

below: The "before" image of the star, from the Sloan Digital Sky Survey [Image courtesy of SDSS]
bottom: The "after" image, from Swift's Ultraviolet / Optical Telescope, where the light from the star explosion outshone the entire host galaxy
[image courtesy of NASA / Swift / UVOT]

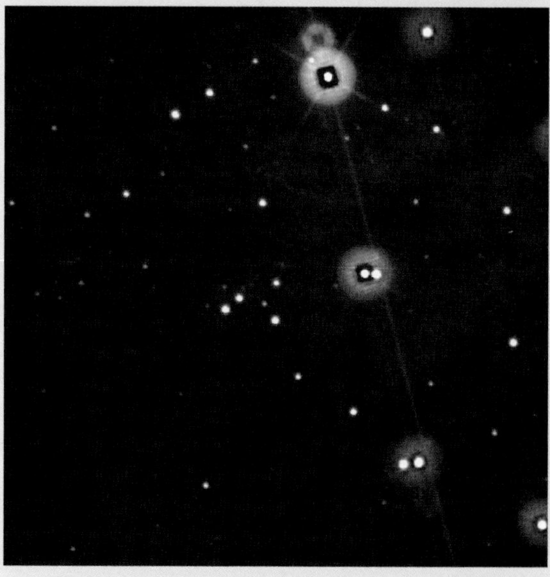

Far, far away

For the first 300 million years during the cosmic "Dark Ages", it is thought no stars shone

Astronomers have detected six of the oldest and most distant galaxies from Earth yet discovered. They are estimated to be more than 13 billion years old and were located using the Keck II telescope, one of the largest in the world.

The light being detected by the telescope on Hawaii was emitted just 500 million years after the Big Bang, said Professor Richard Ellis of the California Institute of Technology in the US, who led the research. It is calculated that the Universe was created 13.66 billion years ago; for the first 300 million years during the cosmic "Dark Ages", it is thought no stars shone.

When announcing the six galaxies in July 2007, Professor Ellis accepted there might be dispute about the findings, but was confident in his team's calculations about the distance and age. The six star-forming galaxies are so distant and faint that they could only be seen through gravitational lensing, in which massive clusters of galaxies magnify the light as it passes through their gravitational fields.

Prior to the discovery of the group of six galaxies, the oldest and most distant galaxy detected was IOK-1, which was identified by a team led by Masanori Iye of Japan's National Astronomical Observatory in Tokyo using the Subaru Telescope on Mauna Kea in Hawaii. The light from the galaxy was produced 12.88 billion years ago and emitted more than eight billion years before Earth was formed.

Another study attempted to establish at what point in the history of the Universe did large galaxies begin to form from smaller galaxies merging. Using the Hubble Space Telescope, researchers from the University of California in the US found hundreds of large galaxies that were born about 900 million years after the Big Bang. They were unable to get any confirmed observations of any created from about 700 million years after the start of the Universe, though they did get one unconfirmed observation.

below: Diagram detailing the gravitational lensing technique. Typical magnification factors are 20 times, thereby bringing into view young galaxies that would otherwise be unobservable [courtesy of Caltech]

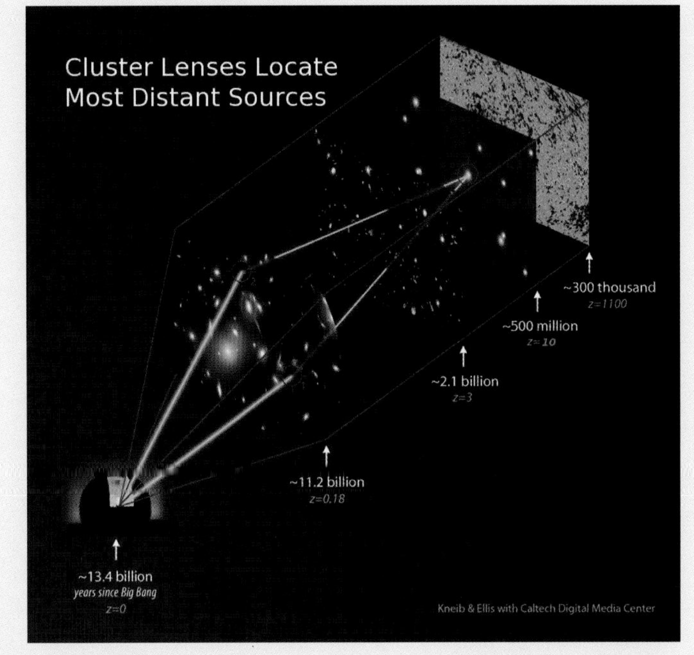

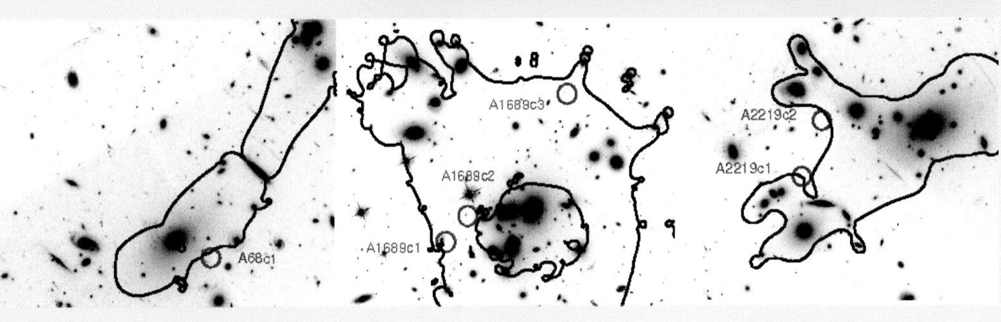

right: A selection of Hubble Space Telescope images of the galaxy clusters with the newly located sources marked in red [courtesy of Caltech]

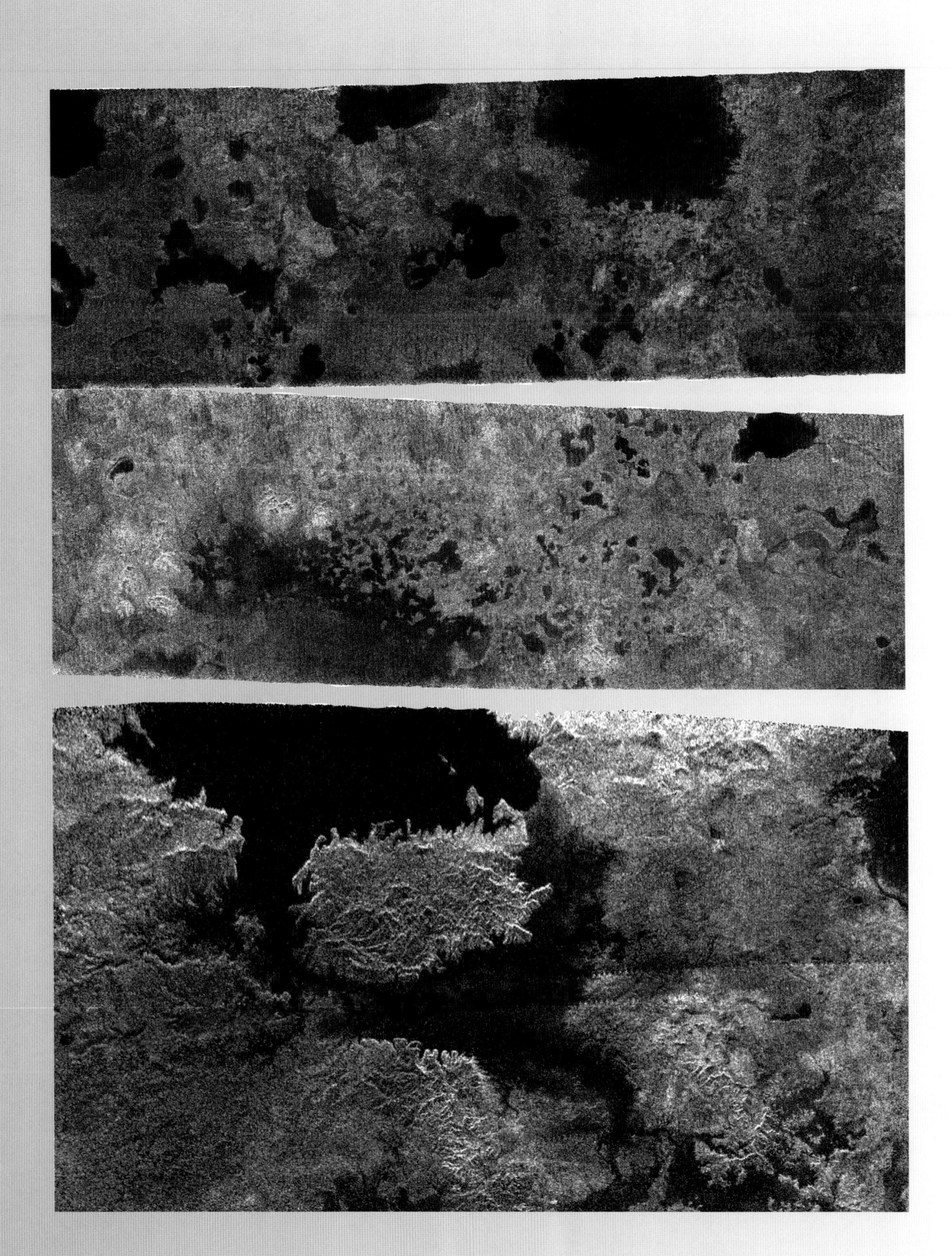

Methane seas

Seas thought to be filled with liquid methane or ethane have been detected on Titan, Saturn's largest moon, providing a hint of the conditions on Earth before life emerged. NASA's Cassini spacecraft found the seas—one of which covers an area of at least 39,000 square miles (101,000 sq km)—along with hundreds of lakes. Radar readings detected dark patches on the surface of Titan, which scientists believe show pools of liquid, though they are subject to confirmatory flypasts and tests. Methane and ethane are the most likely liquids filling the lakes and seas, since they are known to be abundant in the atmosphere while water is absent.

The lakes and seas, the biggest being larger than any lake on Earth, were found close to Titan's north pole, except for one lake located near its south pole. Physical features associated with the dark patches, which look as if they are smooth like bodies of water, support the idea of them being liquid-filled. In particular, the patches seem to be located in depressions, and channels appear to lead to them.

Before Cassini was sent to investigate Saturn and its moons, scientists anticipated finding methane lakes on Titan where, with surface temperatures of minus 179°C (290°F), it is cold enough for methane to exist as a stable liquid. Initial searches in 2005 by the Cassini orbiter and the European Space Agency's Huygens probe found evidence that liquid methane falls as rain and flows sporadically in streams.

Dr Ellen Stofan, of University College London in the UK, led the team that identified the first lakes and said the evidence was "definitive". She said that like lakes on Earth, they are likely to be transparent enough that an astronaut standing on the shore would be able to see pebbles at the bottom of the lake's edge. But it is presumed pointless for the astronaut to take a rod and line—life is extremely unlikely to be found on the moon. Titan's atmosphere is primarily formed of methane and nitrogen, just as Earth's was four billion years ago before the emergence of life, which makes the moon a useful early-Earth model. Given another four billion years, by which time the Sun is expected to have expanded into a red giant, Titan might warm up sufficiently for life to evolve.

The Cassini-Huygens mission to Saturn, its rings and its moons is a joint project between NASA, the European Space Agency and the Italian Space Agency. It took off in 1997 and reached Saturn in 2004, having since sent back a host of pictures and data.

Initial searches by the Cassini orbiter and the Huygens probe found evidence that liquid methane falls as rain and flows sporadically in streams

opposite, top and centre: Images collected by Cassini's radar system showing very strong evidence for hydrocarbon lakes on Titan. The dark patches, which resemble terrestrial lakes, seem to be sprinkled all over the high latitudes surrounding Titan's north pole
opposite, bottom: Radar image showing a big island in the middle of one of the larger lakes imaged on Titan
[images courtesy of NASA / JPL]

Cloud cover

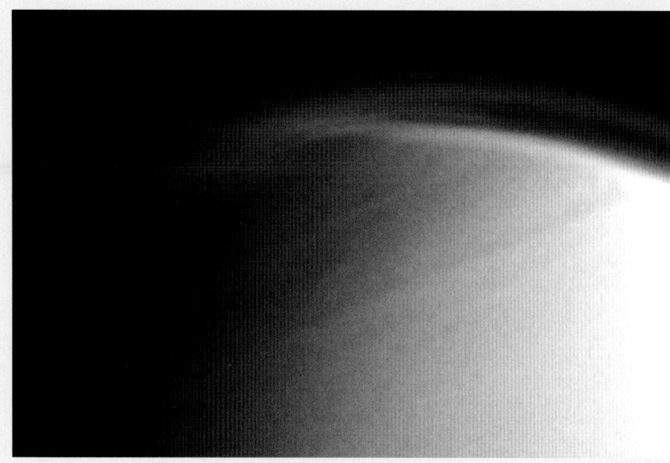

A giant cloud almost 1,500 miles (2,400 km) wide has been pictured covering virtually the whole of Titan's north pole. Cloud cover over the pole had been expected, but its sheer scale and structure took scientists by surprise when images were beamed to Earth. It was pictured using visual and infrared mapping on board Cassini, and with a diameter of 1,490 miles (2,398 km) was big enough to cover half of the United States.

The cloud is made up of methane, ethane and other organic chemicals, strengthening the idea that methane rains down on to the surface to run off into lakes, evaporating once again to form clouds. Clouds on Titan are thought to be part of a cycle in which they are active for 25 Earth years, disappear for four or five years, and then return for another 25 years.

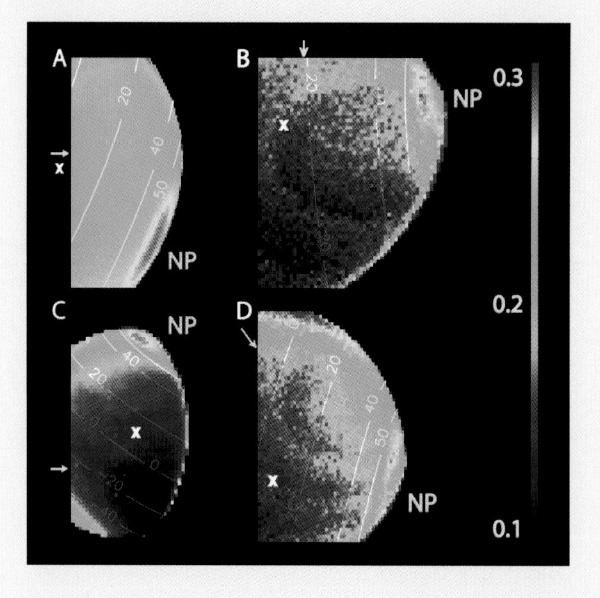

Clouds on Titan are active for 25 Earth years

top left: Image of Titan in ultraviolet and infrared wavelengths taken by Cassini's imaging science subsystem top right: Multiple haze layers near Titan's north pole [images courtesy of NASA / JPL / Space Science Institute]
above: Ethane clouds over Titan recorded by Cassini's visual and infrared mapping spectrometer
[image courtesy of NASA / JPL / University of Arizona]

Biggest star

The biggest star ever recorded was found to have a mass 114 times greater than the Sun. It is in the unromantically named A1 binary star system, 20,000 light years from Earth at the heart of a star cluster in the southern Milky Way. Measurements by astronomers from the Université de Montréal in Canada showed it to be the first star identified with a mass more than 100 times the Sun. Its companion star was not far off, weighing in at 84 times the mass of the Sun to make it the second-most-massive star recorded. The masses were calculated using data picked up by instruments on the Hubble Space Telescope and the Very Large Telescope at the European Southern Observatory in Chile.

It has a mass 114 times greater than the Sun

left: Ultraviolet images of the Sun from NASA's TRACE spacecraft. The relatively cool dark areas have temperatures of thousands of degrees [images courtesy of Stanford-Lockheed Institute for Space Research / NASA]

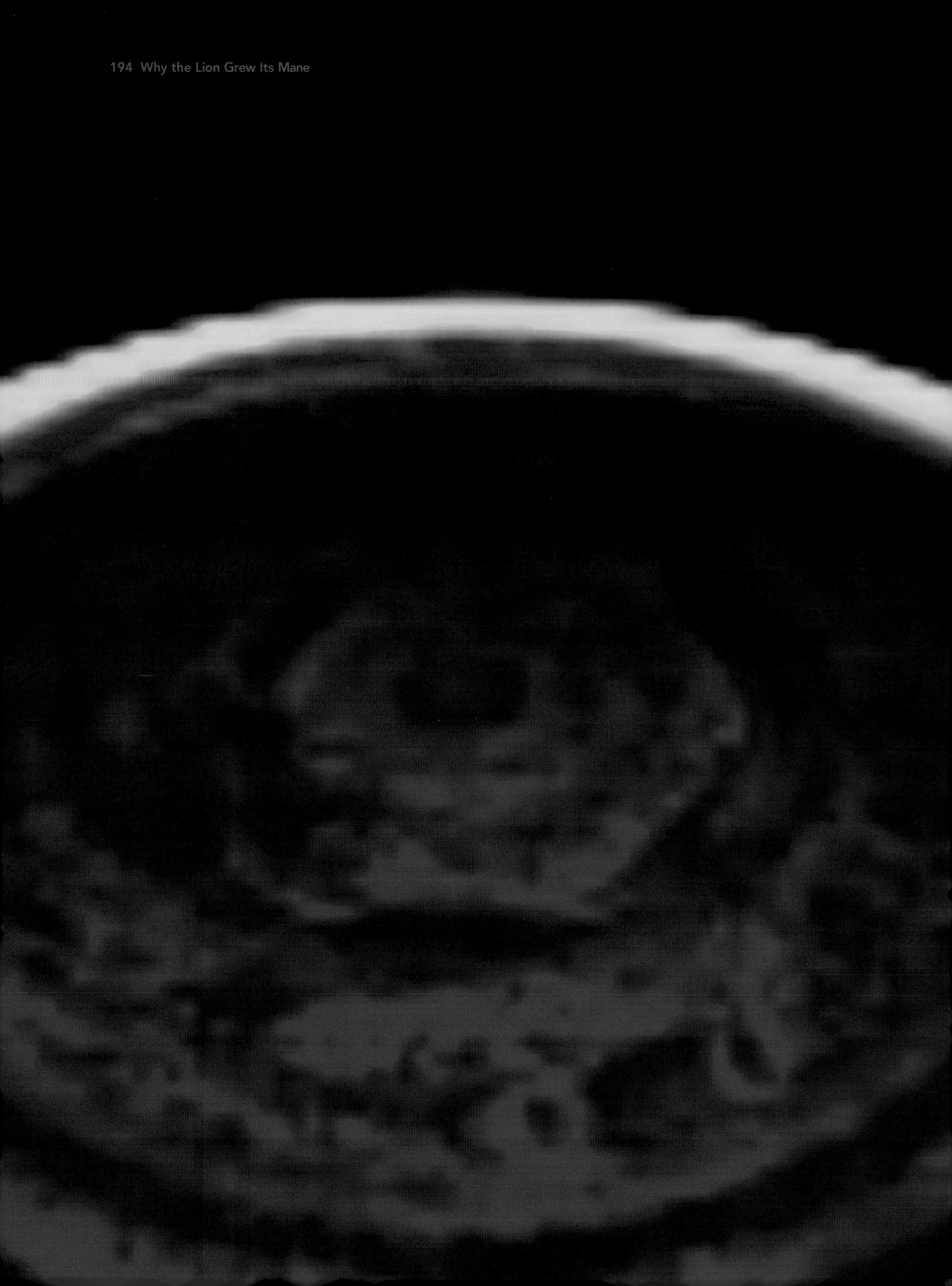

Hexagonal pole

A hexagonal feature in the thick atmosphere of Saturn is among the most intriguing images sent by Cassini. It sits over the planet's north pole and has been seen nowhere else in the solar system. The Voyager 1 and Voyager 2 spacecrafts originally detected the mysterious feature, but Cassini was the first to show it in its entirety in a single picture. Cassini's data reveals that the feature is much deeper than previously realised and extends 60 miles (97 km) beneath the top of the clouds.

The hexagonal feature, with each of its six sides approximately the same length, is huge. At 15,000 miles (24,140 km) across, it is large enough for three Earths to sit within its boundaries. It is similar to the polar vortex on Earth, where winds blow in a circular pattern round the poles, but researchers are still trying to explain why the pattern is hexagonal on Saturn. They have so far discounted auroral activity and radio emissions. Cassini's infrared mapping spectrometer was used over a 12-day period to provide images of the hexagon.

Among other discoveries made through Cassini's data is the likelihood of geological activity on several of Saturn's moons. Geysers shooting out icy particles in plumes stretching about 300 miles (480 km) above the surface were discovered on Enceladus, a moon only 314 miles (505 km) wide. The plumes were ejected from the south pole region, in an area dubbed "Tiger Stripes" because of the pattern of cracks in the surface of the ice-bound moon. Enceladus's south pole was found to be warmer than the rest of the moon and there were fewer craters, suggesting geological activity was smoothing them. Researchers studying the plumes concluded that the stresses created by Saturn's own gravitational tides open cracks 75 miles (120 km) long when Enceladus is furthest from the planet during its 1.3-day orbit, and force them shut again when it is nearest. Dr Terry Hurford, of NASA's Goddard Space Flight Centre, was one of the lead researchers and said it should be possible to predict when each of the stripes will open and erupt.

In another study it was calculated that Enceladus is covered in ice at least three miles (4.8 km) thick, and probably more. The ice shell is thought to hide an ocean. Dr Francis Nimmo of the University of California, Santa Cruz in the US led the study and suggested the heat in the moon's southern regions is generated by ice plates rubbing against each other, much as people warm their hands in winter by rubbing them rapidly back and forth. His team calculated that the faults shift 1.6 feet (0.5 m) during each tidal cycle as Saturn's gravity squeezes the moon.

Titan is also active, and in mid-2007 it was suggested that Tethys and Dione should be added to the list of Saturn's active moons. Cassini identified activity on Tethys and Dione when it detected streams of gas particles being flung into space, possibly by volcanoes. By tracking the gas back, the moons were identified as the source.

At 15,000 miles across, the hexagonal feature is large enough to fit three Earths in its boundaries

opposite: View of Saturn's north pole from NASA's Cassini orbiter reveals the bizarre six-sided hexagonal feature
below: Artist's concept of Cassini during the Saturn Orbit Insertion maneuver, just after the main engine has begun firing
[images courtesy of NASA / JPL / University of Arizona]

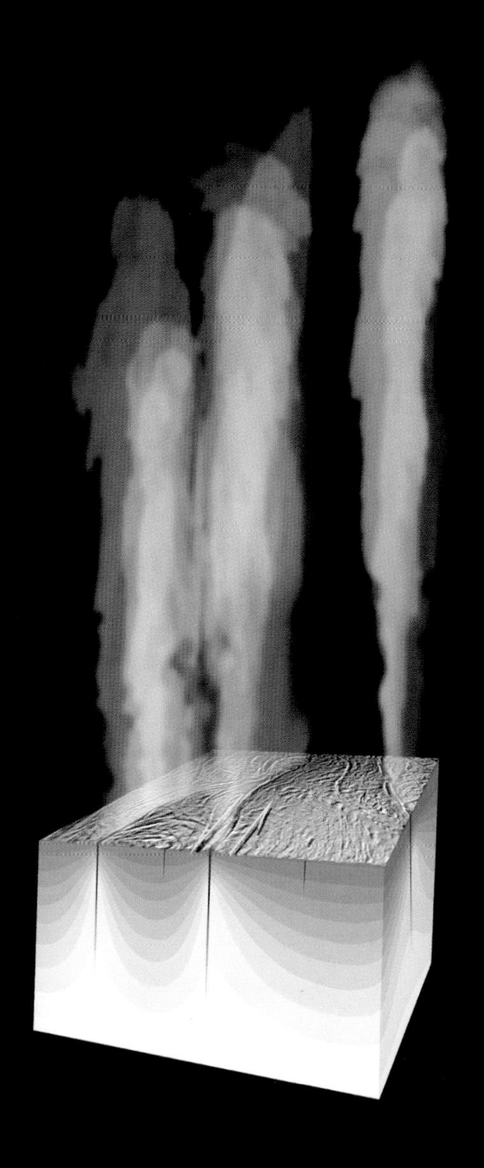

Geysers shooting out icy particles in plumes stretching about 300 miles above the surface were discovered on Enceladus, a moon only 314 miles wide

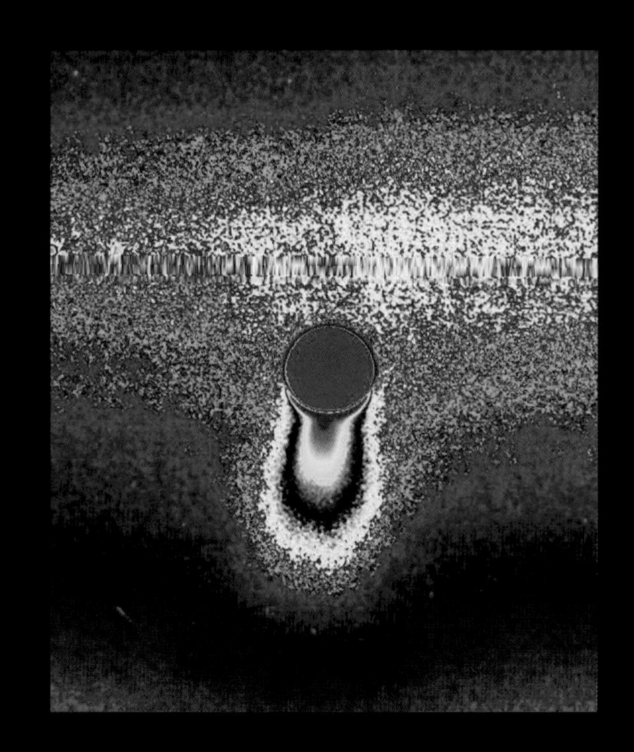

above: Plumes of water vapor and other gases escape at high velocity from the surface of Saturn's moon Enceladus, as shown in this artist concept
[image courtesy of NASA / JPL-Caltech]
right: Ice jets of Enceladus send particles streaming into space hundreds of miles above the south pole
opposite: Saturn's moon Enceladus
[images courtesy of NASA / JPL / Space Science Institute]

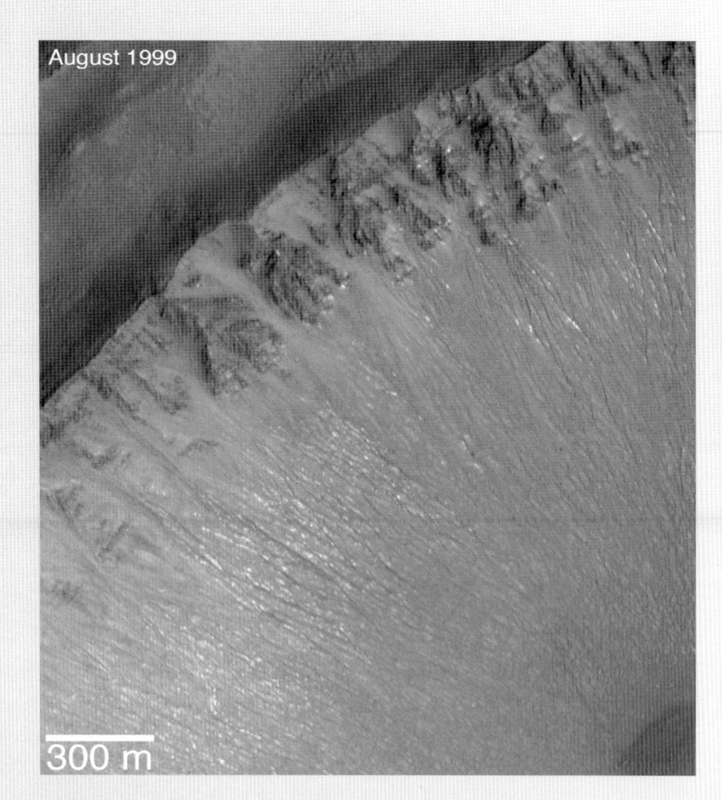
August 1999
300 m

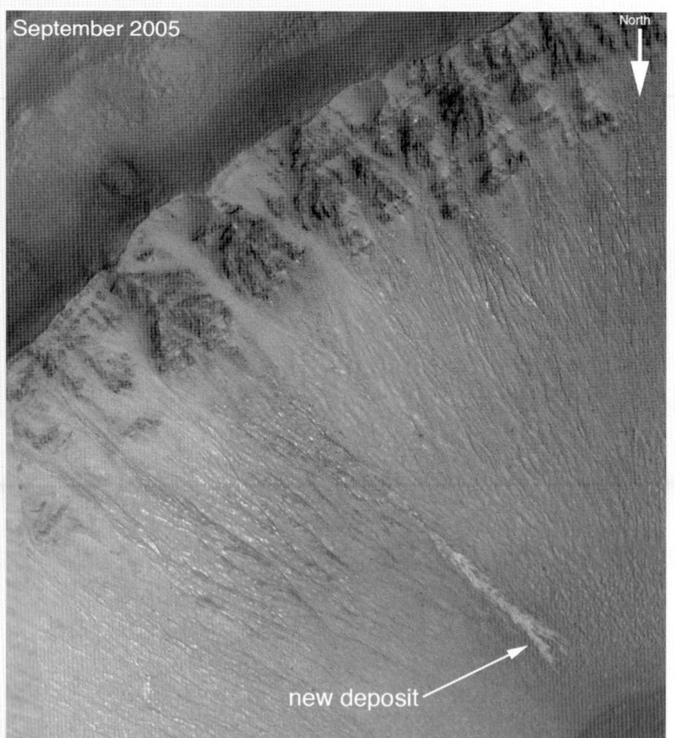
September 2005
North
new deposit

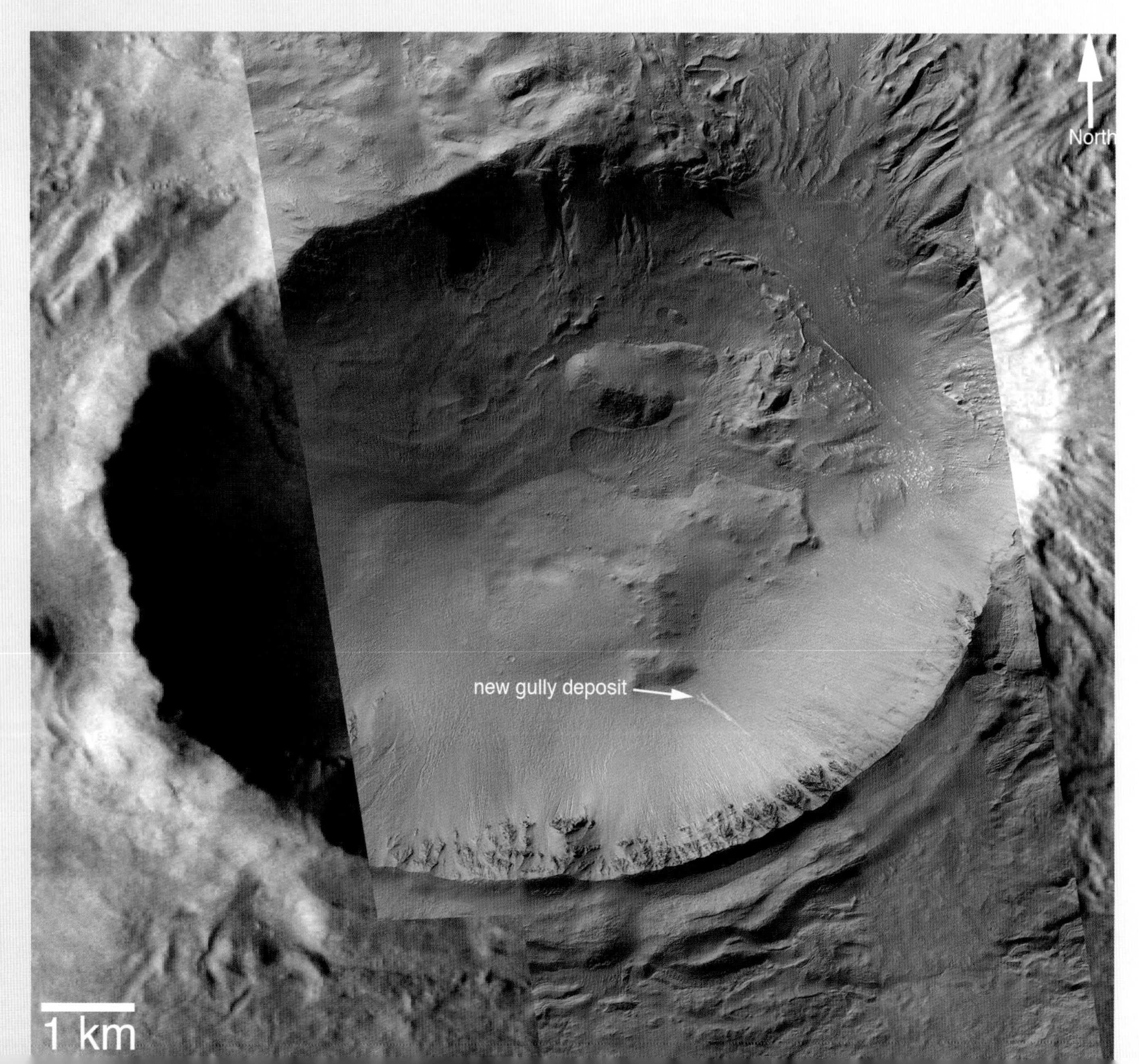
North
new gully deposit
1 km

Water on Mars

Water has been sloshing about the surface of Mars within the last few years, according to analysis of pictures taken by an orbiting spacecraft. Images of the Red Planet taken from 1999 to 2005 showed changes in the surface that scientists are convinced were caused by water pouring through craters. Dr Michael Meyer, of NASA's Mars Exploration Programme, said the photographs "give the strongest evidence to date" that water still flows on the planet. The pictures showed what appeared to be pale-coloured material freshly deposited in two craters. The flow pattern, particularly the way it went around solid obstacles, was highly suggestive of a liquid flowing over the surface.

The deposits were absent from earlier photographs, but appeared by the time the Mars Global Surveyor spacecraft returned to the sites and took pictures of thousands of gullies seen on Martian slopes. Tens of thousands of these gullies had been discovered earlier in the Mars Global Surveyor mission. It was theorised that at least some of them could have been formed by water flows, so the spacecraft was programmed to fly past them again so that "then and now" comparisons could be made.

Photographs of the Terra Sirenum crater were taken in December 2001 and April 2005, while a second crater, unnamed but in the Centauri Montes region, had images recorded in August 1999 and April 2005. At some point between each site being photographed, enough water to fill up to 10 Olympic swimming pools had washed down the crater slopes over a length of several hundred yards. Any water on Mars will either evaporate or freeze rapidly because the planet's atmosphere is too cold and thin for it to remain on the surface in liquid form. Researchers suggested that the water washing down the sides of the craters would have burst from an underground source and was able to flow just briefly, dumping debris before disappearing. The discovery immediately raised hopes that life may yet be detected on Mars, at least in the form of microbial life, which, if supplied with water, could perhaps cope with conditions on the planet.

By photographing the surface at different times, the researchers were also able to assess how often new craters were formed by impacts from meteorites. The orbiter photographed 98 per cent of the planet in 1999 and 30 per cent in 2006, allowing scientists to detect 20 fresh craters in the seven-year period, the biggest being 486 feet (148 m) in diameter. Knowing how frequently craters are formed helps researchers calculate the age of planetary features, those with the fewest impacts being likely to be the most recently formed.

The Mars Global Surveyor orbiter began mapping the Red Planet in 1996 and had notched up 10 years in operation before controllers lost contact in November 2006, probably because of battery failure.

Enough water to fill 10 Olympic-sized swimming pools had washed down the crater slopes

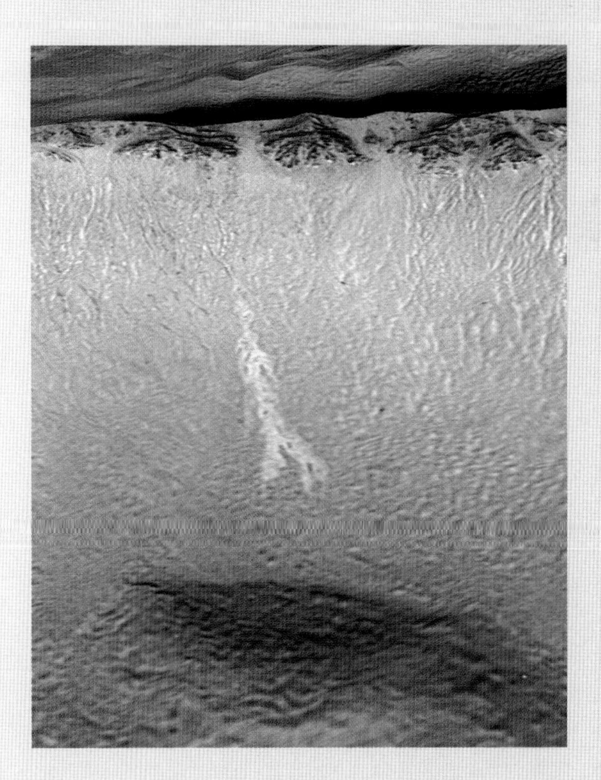

above: Gully deposit in a crater in the Centauri Montes Region
opposite top: The southeast wall of the crater in the Centauri Montes region, as it appeared in August 1999, and later in September 2005.
opposite below: A mosaic of several Mars Global Surveyor images, pointing at the new gully deposit in the unnamed crater
[images courtesy of NASA / JPL / Malin Space Science Systems]

esa

The big freeze

Ice deposits extensive enough to cover the whole of Mars with water to a depth of 36 feet (11 m) have been detected at the south pole of the planet.

The ice was found buried beneath the surface after it was mapped by the European Space Agency's Mars Express orbiter. The discovery was announced in March 2007, just three months after the images suggesting water still flowed were released publicly.

Researchers estimated that the ice detected towards the south pole had a volume of 380,000 cubic miles (1.6 million cu km) and was even greater than the 288,000 cubic miles (1.2 million cu km) of ice already known to be under the northern polar region. The finding helps explain what happened to all the water, which scientists believe was once widespread over the planet.

To locate the hidden ice, the spacecraft used the Mars Advanced Radar for Subsurface and Ionospheric Sounding (MARSIS) instrument. By sending signals to depths of more than 2.3 miles (3.7 km) below the planet's surface, it was able to detect the composition of the subsurface.

Jeffrey Plaut, of NASA's Jet Propulsion Laboratory and one of the lead researchers, said the radar readings provided the most accurate estimates yet seen of the quantities of ice in the southern polar region. In one of the areas surveyed, radar data even suggested a thin layer of liquid water.

Disappointingly for anyone hoping to detect life on Mars, researchers analysing the data felt the temperatures in the ground were too cold to allow for any of the ice to have melted, so considered it more likely there must be some other explanation for the reading.

Data from the Mars Express orbiter suggested a thin layer of liquid water

opposite: The Mars Express orbiter
[illustration courtesy of ESA / D. Ducros]
right: The Mars Express orbiter with its
130-foot (40 m) MARSIS antenna
[illustration courtesy of NASA / JPL-Caltech]

Silica Valley

NASA's Mars Rover Spirit, one of two robotic vehicles exploring the planet, has picked up evidence that water was once on the surface of Mars in great quantities.

One of its six wheels had seized up and, as it was dragged along by the rest of the vehicle, dug a deep furrow into the ground to reveal unusually bright soil. Spirit used its Mini-Thermal Emission Spectrometer to analyse the soil, which turned out to contain high concentrations of silica, a clear indicator that water was once present.

The two known ways that the soil could have formed both require water. One involves water containing dissolved silica being heated by volcanic activity and pushed up to the surface, then evaporating to leave the silica behind. In the other method, hot and highly acidic steam from a volcanic eruption sprays the surface and leaches away everything but the silica on the ground.
Researchers said the find was significant because it provides further evidence to support the theory that Mars was once a much wetter planet. Understanding when, where and in what quantities water was present will help them determine whether life was ever on the planet.

The discovery was made as Spirit explored a depression, now dubbed Silica Valley, in the enormous Gusev Crater. Spirit's twin rover, Opportunity, has been exploring the Victoria Crater and examining its geological features to determine its history. Both rovers were only planned to last three months on the surface of the planet, but their performances have exceeded all expectations since landing on Mars in January 2004.

One of the wheels seized and was dragged by the rest of the vehicle, digging a deep furrow to reveal unusually bright soil

opposite: Artist's concept of the twin Mars Exploration Rovers, Spirit and Opportunity [image courtesy of NASA / JPL-Caltech]
right: The track of disturbed soil, roughly eight inches (20 cm) wide, left by one of the seized wheels of the Mars Rover Spirit [image courtesy of NASA / JPL / Cornell]

Brought down to size

The solar system has been stripped of its ninth planet after astronomers decreed that Pluto failed to satisfy. Amid concerns that Pluto was too small and had too eccentric an orbit, the distant object was eventually downgraded in 2006 on the grounds it lacked the gravitational punch to clear enough debris from its path through space.

Pluto failed to satisfy: it is too small, has too eccentric an orbit and lacks gravitational punch

The International Astronomical Union (IAU), in agreeing a new definition of "planet", determined that Pluto didn't qualify. Instead, the ninth planet was redefined as the first dwarf planet, a new classification in which it is considered the prototype.

One of the key points against Pluto being described as a planet was the discovery in 2005 of an object called 2003 UB313, briefly dubbed Xena after a fictional warrior princess. It orbits the Sun, and with a diameter of 1,491 miles (2,400 km) is slightly bigger than Pluto, which has a diameter of 1,410 miles (2,302 km)—but there was little appetite to define it as the tenth planet.

In case any more controversial rocks are discovered, a planet is now defined as orbiting the Sun, having enough mass to form a sphere, and having cleared space debris in or near its orbit. A dwarf planet is much the same except for its inability to clear away nearby space debris through its gravity. Pluto was joined on the first list of dwarf planets by 2003 UB313 and Ceres, which until then was considered the largest asteroid in the asteroid belt.

Shortly after Pluto was relegated from its status as a planet, object 2003 UB313 was officially named 136199 Eris. The name was taken from the Greek goddess of discord and strife —perhaps appropriately considering the long-running debate over Pluto's status—and suggested by the team that discovered Eris, led by Professor Michael Brown of the California Institute of Technology. They also suggested the dwarf planet's moon should be called Dysnomia, the daughter of Eris. Dysnomia is about 93 miles (150 km) in diameter and is positioned 23,000 miles (37,000 km) from Eris, with the lunar month lasting 16 days.

Any hopes Pluto lovers entertained of winning a reprieve for the former ninth planet suffered a further setback in June 2007, when it was calculated that Eris is 27 per cent more massive than Pluto. Using data from the Hubble Space Telescope and and the Keck Observatory, Professor Brown led research calculating that Eris, which takes 560 years to orbit the Sun, has a mass of 36.6 billion trillion pounds (16.6 billion trillion kg). Pluto is a mere 28.7 billion trillion pounds (13 billion trillion kg).

Eris gets its yellowish colour, researchers believe, from a layer of methane that has seeped from the interior to the surface where it has frozen in temperatures well below 204°C (400°F)—even colder than Pluto.

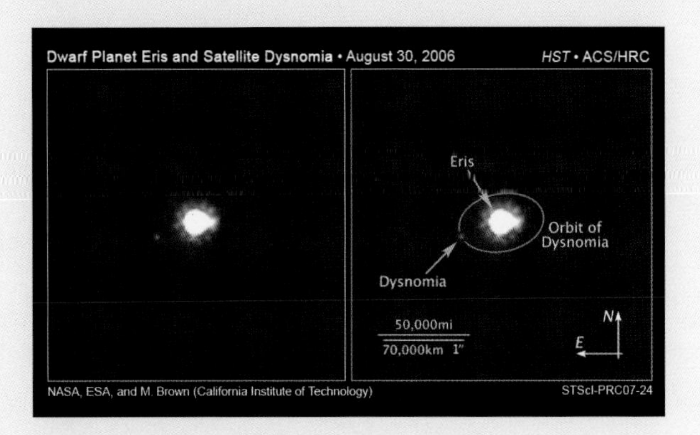

above: Dwarf planet Eris and moon Dysnomia [image courtesy of NASA, ESA & M. Brown (Caltech)]

opposite: An artist's concept shows the Pluto system from the surface of one of its candidate moons. Pluto is the large disk at centre, right. Charon, the system's only confirmed moon, is the smaller disk to the right of Pluto. The other candidate moon is the bright dot on Pluto's far left [image courtesy of NASA, ESA & G. Bacon (STScI)]

Dark matter

A collision of galaxies has provided an unprecedented glimpse of a cosmic structure made from some of the most mysterious matter in the Universe. Dark matter is invisible, but a ring of it was detected five billion light years from Earth. It was discovered by astronomers because of the way it bent the radiation from a cluster of galaxies—just as ripples in water obscure objects at the bottom of ponds when seen from above. It was so unexpected that Dr James Jee, who detected the ring, thought there must have been a glitch in the data received from the Hubble Space Telescope. He spent more than a year trying to make adjustments before realising the ring was as real as the stars in the image. The ghostly ring is 2.6 million light years across and is thought to have formed when two galaxy clusters collided an estimated one to two billion years ago.

The discovery was described by researchers as some of the strongest evidence yet supporting the existence of dark matter, which has puzzled scientists since first being proposed in 1933 by the astronomer Fritz Zwicky. Dark matter is thought to account for 23 per cent of all the mass in the Universe and provide the additional source of gravity required to stop galaxy clusters flying apart. Visible matter, such as stars, is calculated to account for four per cent of the Universe's mass, and the remainder is dark energy—as mysterious as dark matter, but believed to be responsible for the Universe's accelerating expansion. Dark matter's composition is unknown but is held to be a particle common across the Universe, with countless numbers passing through everyone.

Dr Jee, of Johns Hopkins University in the US, led research concluding that when the galaxy clusters smashed into each other in a "titanic collision", the dark matter expanded from the centre of the cosmic crash. As it moved outwards it slowed down as gravity exerted its pull, with the dark matter at the back pushing up against the slower dark matter at the front to create the ring, much like traffic builds up behind tractors on narrow roads.

The ghostly ring is 2.6 million light years across and is thought to have formed when two galaxy clusters collided an estimated one to two billion years ago

left: This Hubble Space Telescope composite image shows a ghostly "ring" of dark matter in the galaxy cluster Cl 0024+17 [image courtesy of NASA, ESA, M. J. Jee & H. Ford (Johns Hopkins University)]

Double big bang

A star so massive that it blew up twice during its death throes has been detected for the first time. Explosions from the huge star, estimated to be 60 to 100 times bigger than the Sun, were observed in 2004 and again in 2006. Koichi Itagaki, a Japanese amateur astronomer who specialises in finding supernovas, first detected the two explosions, both of which emanated from the same region of space 78 million light years from Earth.

Professor Stephen Smartt and Dr Andrea Pastorello from Queens University, Belfast, in the UK, realised the explosions came from the same region of space in galaxy UGC 4904 and had to be from the same star. They teamed up with Mr Itagaki and French, Italian and Chinese scientists to analyse the data from when the explosions were first detected, and confirmed the double explosions were from one rather than two stars. The research team concluded that what had been seen was the death of one of the biggest stars in the Universe, and that it probably became a black hole after tearing itself apart. It is likely that the first observed explosion, possibly one of a series taking place, was a result of the star's outer atmosphere being blown away, which was followed by a second and bigger explosion when it became a supernova.

The star, estimated to be up to 100 times bigger than the Sun, became a supernova after blowing up twice during its death throes

right: Swift X-ray Telescope image of Supernova 2006jc in the galaxy UGC 4904
far right: Swift Ultraviolet / Optical Telescope image of Supernova 2006jc [images courtesy of NASA/Swift/ S. Immler]

Explosive Moon

The Moon was geologically active for hundreds of millions of years longer than previously believed and may still have volcanoes, observations have shown. Evidence of a gas volcano erupting just two million years ago was found in a depression 1.7 miles (2.7 km) wide, shaped like the heel of a shoe. Rather than magma, huge quantities of gas would have erupted violently from the volcano, hurling aside surface rocks and exposing a long-buried layer of basalt. Rocks on the surface of the depression are in such good condition that planetary geologists calculated they could only have been exposed for one to 10 million years, with two million the most likely. It was previously thought that the last volcanic eruption on the Moon took place about a billion years ago.

The depression, called Ina, is at the peak of a dome 1,000 feet (300 m) high, and lies at the intersection of valleys or rilles, which on Earth are often the locations for geological activity. Four other potentially active sites have also been identified. One of the lead researchers, Professor Peter Schultz of Brown University in the US, said the sharpness of the rock features showed up in pictures taken during the Apollo missions. It is expected that space debris hammering into the region would have destroyed the freshness of the rock surfaces long ago, certainly within 50 million years, had the eruption not taken place in the Moon's recent past. He said such features could still be forming today and that huge eruptions of gas could explain why amateur astronomers have observed puffs or flashes of light from the Moon's surface.

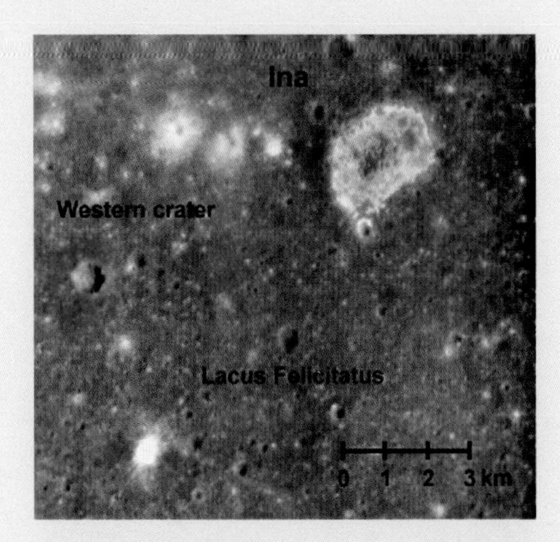

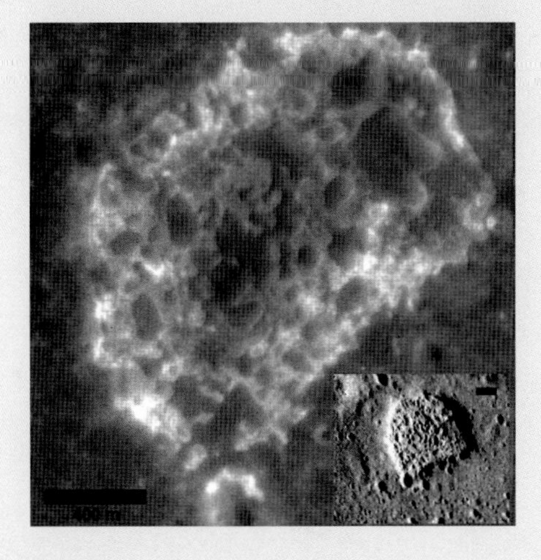

top left: A false-color composite photo of Ina and a nearby young crater. Blue denotes freshly-exposed titanium basalts, while green traces immature (relatively unweathered) soils.
left: The strange-looking geological feature named Ina, shaped like the letter "D" and about two miles wide
[images courtesy of NASA]

Huge quantities of gas would have erupted violently from the volcano, hurling aside surface rocks and exposing a long-buried layer of basalt

Colour of life

Astronomers searching for life on other planets should avoid the colour blue and instead keep their eyes open for red, green and yellow. Blue is the colour most likely to be useful to plants on extrasolar planets because it is the most energetic, according to astrobiologists. Red, yellow and green are the colours likely to be of least use to the plants that are hoped to be covering wide swathes of other planets.

Plants will reflect back whichever colour of light turns out to be the most useless to photosynthesising, just as green is rejected by most vegetation on Earth. Chlorophyll in the Earth's plants absorbs more blue and red light than green.

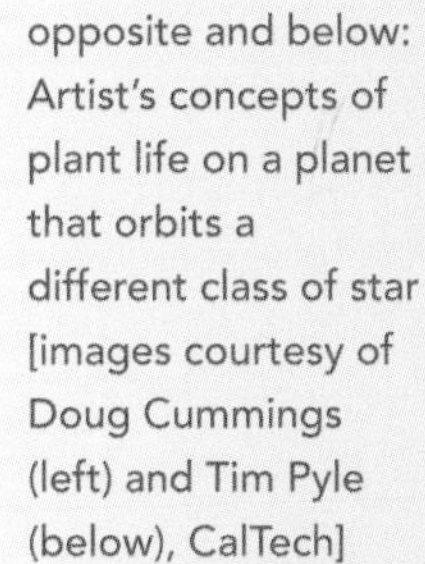

opposite and below: Artist's concepts of plant life on a planet that orbits a different class of star [images courtesy of Doug Cummings (left) and Tim Pyle (below), CalTech]

Dr Nancy Kiang of NASA's Goddard Institute of Space Studies said all the colours of the rainbow are likely to be reflected by extraterrestrial plant life, but blue will be bounced back the least because of its value as a source of energy. The findings of two studies by NASA and the California Institute of Technology into the use of light by plants are intended to guide the searches for Earth-like planets in other solar systems.

The search for potentially habitable planets is expected to accelerate in the next few years with the planned launch of NASA's Terrestrial Planet Finder telescope and the orbiting telescopes of the Darwin Flotilla, proposed for 2015 by the European Space Agency.

Tiazac
300

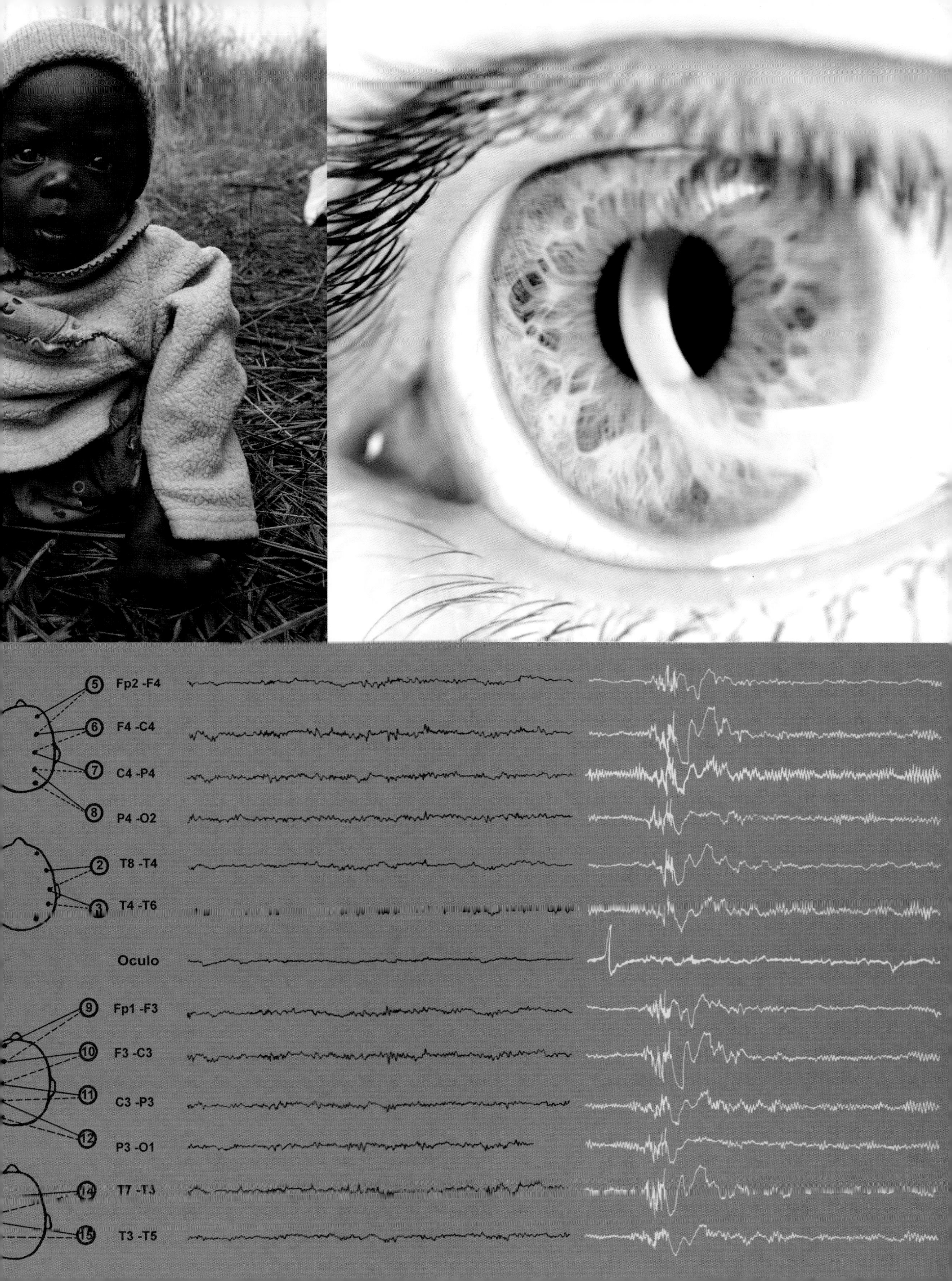

Fp2 -F4
F4 -C4
C4 -P4
P4 -O2
T8 -T4
T4 -T6
Oculo
Fp1 -F3
F3 -C3
C3 -P3
P3 -O1
T7 -T3
T3 -T5

previous, clockwise from top left: Father-daughter bonding [image: © iStockPhoto / Ronald Bloom], child in eastern Congo [photo courtesy of WFP / Susan Schulman], human eye [photo: © iStockPhoto / Hirlesteanu Constantin-Ciprian], brain waves during sleep [image: Science Photo Library], Damien Hirst *Pharmaceuticals* [photo: © Damien Hirst], George Clooney as Dr Doug Ross on the television series ER [photo: Getty / Handout]

What's in Our Heads

Thinking is a part of our being that goes unnoticed most of the time, yet we couldn't do without it. Everything we do—whether wagging a finger or formulating a new rule of physics—requires a degree of mental activity.

Our minds determine who we are. They dictate our personalities, hold our memories, allow us to learn and control physical movement. Yet, for all our reliance on the brain and our recognition that it above all other physical attributes is what allows mankind to so dominate and control the world around him, it is far from fully understood.

Scientists trying to unlock the secrets of the mind are hampered by the fragility of the brain and the lack of physical form of much of what they are trying to study, such as emotions. Nevertheless, researchers in many fields are making significant progress in identifying how the brain functions and gaining insights into what defines us as human.

First impressions

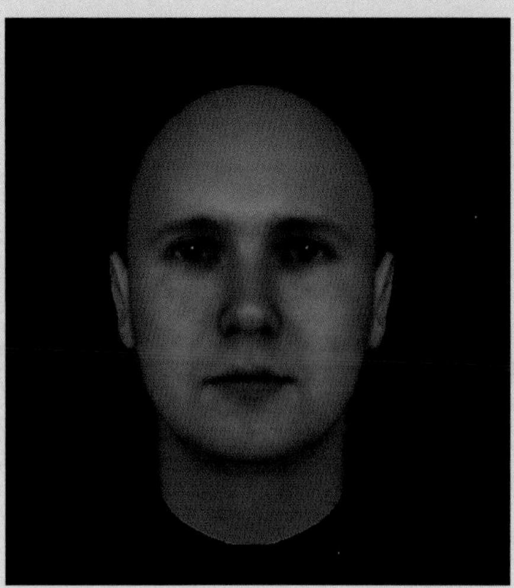

above: Participants in the study were asked to judge composites of a trustworthy face (right) and an untrustworthy face (left) [images courtesy of Alexander Todorov, Social cognition and social neuroscience lab, Princeton University]

Snap judgements might be a hangover from the brain's early evolution in response to fear

Humans make character assessments in the blink of an eye after just a glimpse of a stranger's face, researchers have found. These snap judgements are not necessarily accurate, but they are nevertheless made on whether someone is perceived as likeable, competent or even trustworthy. Dr Alexander Todorov of Princeton University in the US carried out a series of experiments to test the speed of judgements. He found that volunteers came to the same conclusions about a face whether they were given a second, half a second or even just a tenth of a second. Making their minds up so quickly meant the volunteers often had no time to begin rational thought, let alone make a reasoned and considered judgement.

The 200 volunteers that took part in the study were asked to look at pictures of 66 different faces flashing on a screen for a tenth, half or a full second. They reported how trustworthy they felt each of the faces to be and indicated how certain they were in their judgement. Other personality traits assessed included likeability and competence. The length of time allowed for looking at the faces made no difference to the assessment made of the other person's character, but there was a correlation between the volunteer's level of confidence in the judgement and having a few extra tenths of a second to decide.

The reason why snap judgements on personality are made remains unclear, but Dr Todorov suggested it might be a hangover from the brain's early evolution. It appeared, said the psychologist, that the part of the brain involved in fear response could play a role in instant character assessments. This part of the brain, the amygdala, existed long before the prefrontal cortex, which is more involved in rational thought and might, Dr Todorov suggested, be completely bypassed when first impressions are made.

Hidden messages

Subliminal messages have been shown for the first time to be picked up by the brain, as long as it is not busy. Researchers equipped with an fMRI scanner established that what is received by the eye can be registered by the brain even if there is a lack of conscious awareness of what has been seen. The finding supports the often-disputed notion that viewers unconsciously absorb messages flashed up on cinema or television screens for a fraction of a second. Volunteers were fitted with special glasses, through which one lens provided faint images of common household objects such as an iron, and the other created flashes that suppressed conscious recognition of the picture. Monitoring the volunteers with the scanner showed that the pictures of the household objects sparked brain activity despite the subjects of the experiments being unaware of what they had seen.

Further tests carried out by a team led by Dr Bahador Bahrami of University College London in the UK revealed that when engaged in tackling an awkward visual task, the brain completely ignored the subliminal messages. Dr Bahrami said the study hinted at the potential power of subliminal advertising to get messages across to consumers secretly. Whether it would influence them enough to make them want to buy the product, however, remains uncertain.

Blue eye in macro [image: © iStockPhoto / Hirlesteanu Constantin-Ciprian]

A best friend is a woman

If it's friendship that's wanted, it is far better to turn to a woman than a man, according to a study of adult behaviour. Where women are loyal, dependable and true, men are fickle creatures whose reliability is, at the least, questionable.

Behaviour exhibited by more than 10,000 people over a decade was analysed for the study, conducted by researchers from the University of Manchester in the UK. Research concluded that women are more likely to remain loyal to the same friends whatever else is happening in their lives, such as moving long distance to a different part of the country. They display genuine interest in the welfare of their friends, what is happening in their lives and how their families are getting on.

In contrast, when men choose their friends, an element of "what's in it for me" selfishness tends to creep into the equation. Their friends are most likely to be found in bars, where they share a mutual appreciation of a convivial drink. Whereas women try to keep in contact with old friends after moving to other parts of the country, men are quite happy to start afresh with a new circle of friends. Dr Gindo Tampubolon led the research and concluded that while both genders display selfishness and generosity of spirit, friendship for women is much deeper and more altruistic than it is for men.

Women are loyal and dependable, while men are fickle creatures whose reliability as a friend is questionable

left: According to research, female friends are more likely than males to stay true to each other, no matter the situation [photo: Getty]

Forgetting to smoke

A region of the brain linked to the sensations of hunger and pain has been identified as a potential key to getting smokers to quit cigarettes with ease. The insula was pinpointed as such after a man who smoked 40 cigarettes a day was able to give up overnight after suffering brain damage during a stroke. The stroke caused damage to the insula region of the patient's brain and it is thought to be responsible for his ability to give up smoking effortlessly. The patient told doctors that he was able to give up smoking easily because the urge to have a cigarette disappeared after the stroke.

Researchers decided to look for more stroke patients who had suffered damage to the insula to discover whether they had been able to give up smoking easily. Using a patient database held by the University of Iowa in the US, they identified 19 smokers whose insula had been impaired—12 had given up smoking without trouble and one had given up with some effort. The link found between the insula and smoking addiction will, the researchers hope, aid scientists in developing new ways of getting people to stop smoking. In particular, they believe drugs could be developed to target the insula and block cravings.

Surgery and brain implants could eventually be a possibility, but much more still needs to be learned about the insula. It is believed to have a variety of roles, particularly in cravings and emotions, and care would need to be taken to avoid knocking out other functions while aiming to stop an addiction.

The study is thought to be the first of its kind to look at the effect of brain damage on drug addiction. Lead researcher Dr Antoine Bechara of the University of Southern California said one of the implications of the finding is that all addictions, from overeating to heroin abuse, could be affected in the same way by brain lesions in the insula. Six of the 19 stroke patients studied continued to smoke, and the research team suggested this might indicate they suffered damage to a slightly different part of the insula than the others.

A man who smoked 40 cigarettes a day was able to give up overnight after suffering brain damage during a stroke

[image: © iStockPhoto / Slobo Mitic]

Tortured genius

A gene that contributes to mankind's unique intelligence may equally be a factor in mental illness associated with delusions and hallucinations. For most people, a common variant of the DARPP-32 gene makes the brain better at processing information, but at the same time appears to increase the risk of schizophrenia. Researchers found that the gene controls a circuit linked to the prefrontal cortex and affects functions such as intelligence, motivation and attention. While in most cases the variant would be an evolutionary advantage, when other genes and environmental factors disturb the prefrontal cortex—where thoughts and actions are managed—it could assist the onset of schizophrenia.

Scientists at the National Institute of Mental Health in the US carried out the study into the gene variant, for which DNA samples were taken from more than 1,000 people. They established that at least three-quarters of the volunteers carried the variant and that it was more prevalent among people with schizophrenia.

above: Sylvia Plath, revered for her poetry and prose, but also known for her battles with mental illness and eventual suicide

A common variant of the DARPP-32 gene contributes to man's unique intelligence, but may also increase the risk of schizophrenia

Four-legged ducks

The duck was drawn by a woman with a form of dementia that gradually removes the ability to conceptualise

A picture of a duck with four legs has helped solve a 150-year-old argument about which part of the brain deals with concepts. The duck was drawn by a woman with a form of dementia that gradually removes the ability to conceptualise.

The 55-year-old woman was asked to draw a picture of a duck and was quite capable when she had a photograph in front of her. However, when she was asked to draw from memory rather than copy from a picture, she began to struggle because, researchers explained, she needed to be able to recall the meaning of what she had seen and grasp what a duck is.

After a 10-second delay between seeing the photograph and drawing the duck, she began to draw a third leg, but scribbled it out upon realising her mistake. When there was a full minute's delay between seeing the photograph and starting to sketch, the end result was a duck with four legs and a tail like a turkey's.

The test was one of several carried out by researchers from the University of Manchester in the UK to solve the question of which section of the brain is used to understand concepts, words and meaning.

Semantic dementia is a form of the degenerative disease whereby patients become increasingly incapable of recognising, naming and understanding common objects. Professor Matthew Lambon Ralph led the team that tried to identify the part of the brain associated with the dementia and said the duck pictures illustrated the loss of function.

Other tests on patients with the condition included brain scans that revealed they had suffered tissue loss in the anterior temporal lobe, located beneath the ear.

Volunteers helped confirm the idea that the section of brain that stores meanings is the anterior temporal lobe rather than, as previously believed, a part called Wernicke's area. They agreed to take part in experiments where magnetic pulses were used, in a technique called transcranial magnetic stimulation, to tire out their temporal lobes; the effects lasted only a few minutes. When asked questions after undergoing the procedure, they were found to be 10 per cent slower in being able to recall the names of objects they were shown.

below: Three duck drawings by a woman with dementia. The first, far left, was drawn with no delay between being shown a photograph and drawing. The sketches get less accurate with a 10-second delay (middle) and a 60-second delay (right) [images courtesy of Matthew A. Lambon Ralph, Neuroscience and Aphasia Research Unit, University of Manchester]

Use it or lose it

People who are mentally active in old age are 2.6 times less likely to develop dementia than those who let their minds idle. Keeping the mind active by stimulating it with activities such as reading newspapers or playing chess seemed to help ward off the onset of dementia and Alzheimer's disease.

In a study of 700 elderly people with an average age of 80, researchers found that those people who kept their minds active had a significantly reduced risk of dementia over the five-year research period. They calculated the risk as 2.6 times lower in comparison with the mentally inactive as part of a wider project into aging at the Rush University Medical Centre in the US. Researchers found that the risk reduction was present irrespective of the elderly person's mental activity when younger, socioeconomic history and levels of physical exercise. The study team hoped the discovery would help in the search for treatments to prevent Alzheimer's.

It was also discovered that mental activities including reading a newspaper, visiting a library or going to the theatre were associated with the avoidance of mild cognitive impairment, which precedes dementia.

Mental activities including reading a newspaper or going to the theatre were associated with the avoidance of mild cognitive impairment

left and below: Senile plaques seen in the cerebral cortex in a patient with early-stage Alzheimer's disease

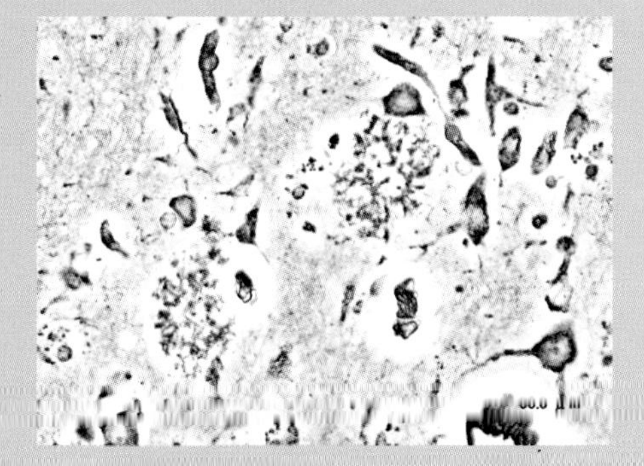

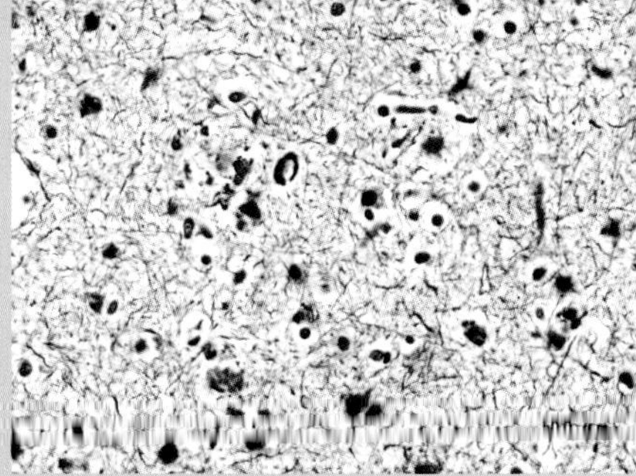

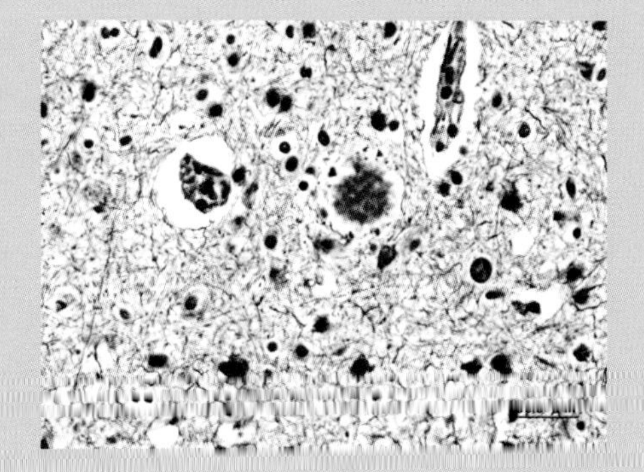

Daddy's girls

A strong emotional bond between father and daughter significantly increases the chances that, when she grows up, the girl will want to find someone just like daddy

left: Forming a father-daughter bond at a young age may influence a girl's choice of partner later in life
[image: © iStockPhoto / Ronald Bloom]

Women who were daddy's girls show a distinct tendency to marry men who look like their fathers, a study has shown. A strong emotional bond between father and daughter significantly increases the chances that when she grows up, the girl will want to find someone just like daddy.

The happy relationship between man and girl seems to imprint in the daughter's mind the idea that the father's facial features are an ideal physical form in a man. For daughters who either get on only reasonably well with their fathers or don't like them at all, there is no correlation between what daddy looks like and what they find attractive as adults.

The study was conducted by researchers from the University of Wroclaw in Poland, the Polish Academy of Sciences, and the University of Durham in the UK. To reach their conclusions, they took a group of 49 Polish women, all of whom were the eldest daughters in their families, and showed them a series of pictures of 15 male faces.

Each woman was asked to select which of the 15 faces she found the most attractive. Ears, hair, neck, shoulders and clothing were hidden from view so that only the facial features were judged. Measurements of the men's noses, chins, eyebrows and other facial features were taken and then compared to similar measurements of the father's faces. The daughters in the study also answered questions about their relationships with their fathers, such as how much time they spent with him as a child and how much love and affection he showed.

When taken as a whole, there was no link between a father's looks and what a daughter found attractive, but when the women who were closest to their fathers were considered separately, there was a clear similarity. Dr Lynda Boothroyd of Durham University said the results showed that where sexual imprinting is concerned, it is quality rather than quantity that counts. Rather than build up an ideal of attractiveness from all the faces around, she said, human brains construct their preferred features from positive influences.

Until a few years ago, the parental role in children's selection of partners later in life was thought to be passive, but a growing body of evidence now suggests it is an important factor in sexual imprinting. The study did not analyse boys, their relationships with their mothers, and the women they choose as partners, but the researchers suspected the same patterns of selection would be evident for them as for fathers and daughters.

Compassion fatigue

Humans are at their most generous and compassionate when there is just one person in need, a psychology professor has found. Images of hundreds or thousands of people in distress and in need of help is likely, tests showed, to invoke a numbing effect that makes individuals less willing to make any effort on their behalf. But show people a single person in need and they are likely to go to great lengths to try to help out.

Professor Paul Slovic of the University of Oregon in the US maintained that the human mind developed the ability to empathise with others—but only for limited numbers. Facing misery and death on a large scale tends to induce "psychic numbing" that reduces the ability to respond compassionately. He carried out experiments to test people's reactions to the suffering of others and found that responses weaken immediately when asked to help more than one individual. Even when people are asked to give money to two children in need rather than one, the willingness to donate falls away.

Professor Slovic said it is likely that the explanation to so-called psychic numbing lies in the circumstances in which the human brain evolved. Empathy and sympathy would have developed, he said, in times when early humans would have needed to worry about at the most a handful of people around them, not during an era of mass communication and a world filled with billions of people. Because the mind was adapted to react to what was going on in small, isolated communities, there would have been no survival value in learning to cope with large-scale disasters.

Empathy and sympathy would have developed in times when the most early humans would have needed to worry about were a handful of people around them

below: People were found to be more willing to help one child in need over aiding two [photos courtesy of WFP]

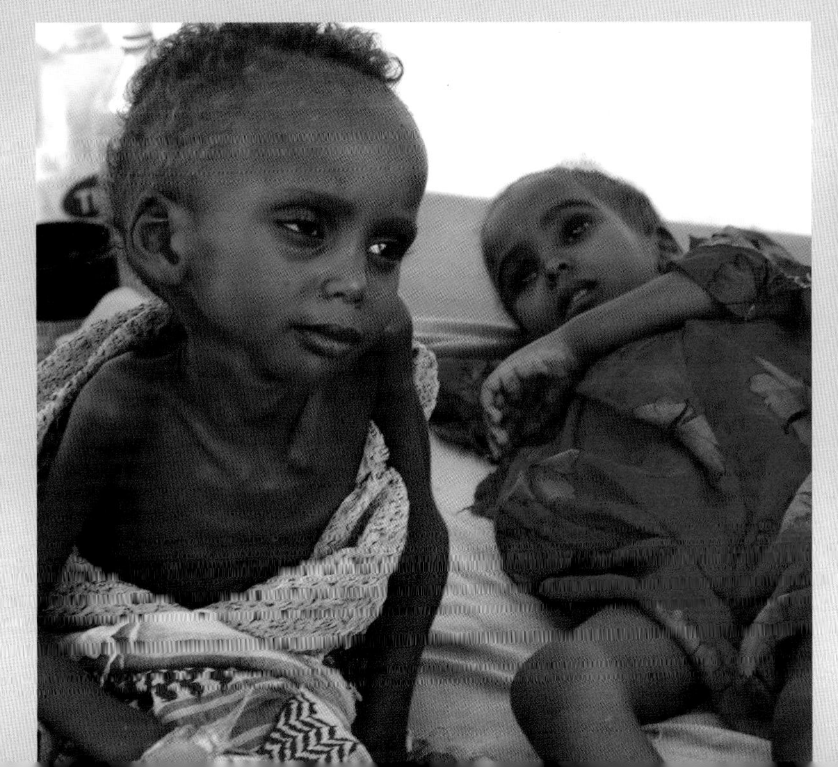

Confidence

cures

The power of placebos to result in a cure has been linked to the brain's release of natural painkillers. Placebos have frequently been found by doctors to lead to an improvement in a patient's condition, but the reason has mystified scientists.

A study in Germany has suggested that when a patient is confident useful medicine has been applied, the brain produces endorphins that result in the patient feeling better—irrespective of the effectiveness of the medicine itself.

Professor Christian Büchel of the University of Hamburg led research in which 19 volunteers were treated while their brains were monitored with functional magnetic resonance imaging. Participants were given laser pinpricks to their hands and told that a pain-relief cream had been rubbed in on one and a placebo cream on the other. In reality, both hands received placebos.

Professor Büchel said that when patients believed they had been given an effective pain reliever, the brain was found to have become more active in the rostral anterior cingulate cortex (rACC), a region of the brain that manages pain control. Areas of the brain linked to the perception of pain were less active, and patients also reported less pain.

The rACC region is known to have high numbers of opiate receptors and the professor said the placebos worked in a similar fashion to morphine or codeine. The findings have provided insights into why it is that many patients who start on courses of antidepressants report improvements in their condition well before the drugs could have taken effect.

Placebos activated the rACC section of the brain when patients believed they were given real medicine

People have confidence in medicine. I noticed they were looking at shiny colours and bright shapes and nice white coats and cleanliness and they were going, right—this is going to be my saviour, except they weren't reading the side effects

Damien Hirst

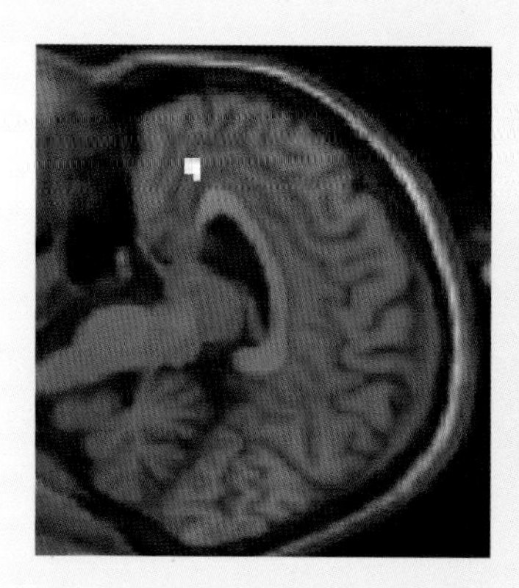

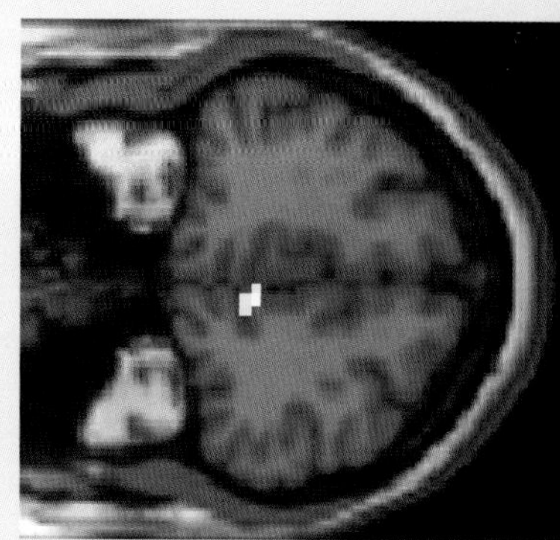

opposite: Damien Hirst, *Pharmaceuticals*, 2005 [photo: © Damien Hirst]
above: Images showing the placebo-related activation of the rACC [courtesy of Dr Christian Büchel, Institut für Systemische Neurowissenschaften]

Oh no, not again

Being accident-prone is not so much simple bad luck as it is a matter of personality

above: Laurel & Hardy [photo: Michael Ochs Archives / Stringer / Getty]

Just as every classroom has its bright kid, its joker and its sporty-type, so it is likely to have its accident-prone child. One in every 29 people, according to a study in the Netherlands, are 50 per cent more likely than the rest to have accidents. The research quantifies for the first time an inherent tendency to fall off ladders, trip over kerbs and slip with the kitchen knife. It shows that having more accidents than others around you is not so much simple bad luck as it is a matter of personality.

For the research, carried out by a team from the University Medical Centre Groningen, data covering 147,000 people in 79 studies was assessed. By analysing the distribution of accidents across the general population, it was found that there were clearly individuals who suffered more accidents than others. The difference was so great that it was concluded it could not be put down to chance.

By arguing against bad luck and for personality, the researchers hoped to end an argument that has rumbled on in scientific circles since 1919, when there was research into accidents at a munitions factory. That early study highlighted the observation that a small number of workers in the munitions factory that supplied British forces in the World War I were responsible for, or victims of, a disproportionate number of accidents.

The Groningen study looked at people from many walks of life, including bus drivers and aircraft pilots. Children, unsurprisingly, were found to be more likely to be hurt than adults, and boys to be more likely than girls to have accidents. But they cautioned that being accident-prone is not necessarily a condition children will naturally grow out of—they may be stuck with the tendency for life. While they were able to quantify the chances of being accident-prone, the researchers were unable to predict which people are most likely to suffer the problem.

Once it is possible to point to who is the most accident-prone it is likely, they said, that insurance premiums will be increased to take it into account. Equally, workplace health and safety could become an issue with the prospect of employers refusing to have accident-prone workers anywhere near dangerous machinery or chemicals.

Heart over head

For the first time, emotion has been demonstrated to play a part in judging between right and wrong. In finding part of the brain that allows emotion to influence, if not rule, the head, researchers identified part of what defines us as humans. The study pinpointed the ventromedial prefrontal cortex (VMPC) as the region of the brain that plays a crucial role in solving extreme moral problems. They found that at the same time as the brain uses logic to solve dilemmas, intuition and emotion also play a role. It was previously thought that emotion was excluded until after a decision had been reached.

It had previously been thought that emotion was excluded until after a decision had been reached

Researchers came to their conclusions after studying 30 volunteers' responses to moral dilemmas, six of whom had damage to the VMPC. Scenarios put to the volunteers included variations on the classic moral dilemma of whether it would be right or wrong to kill an innocent person to save others. Rationally speaking, the answer has to be that one must die to save the rest of the group—but because emotion is involved, most people find reaching a decision less straightforward and are hard pushed to accept the logical solution because they empathise with the victim.

Those volunteers with a damaged prefrontal cortex found it much easier to come to a conclusion because they had a reduced ability to display empathy or compassion, and thus relied more heavily on logic. The six individuals offered strictly rational judgements, whereas the remainder of the study group showed a far greater degree of aversion, or even downright refusal, to killing an innocent despite it being for the greater good.

The research was led by Professor Daniel Tranel, of the University of Iowa in the US, who said two reasoning processes were being carried out simultaneously—one emotional and intuitive and the other cold and rational. People who had lost the use of the VMPC could only think rationally. Dr Antonia Damasio, of the University of Southern California in the US, took part in the study and said she suspected that the brain's inclusion of emotion in decision-making was a response to an accumulation of wisdom over an evolutionary timescale. She said it appeared from the study that the VMPC contributed to "our wisdom and humanity". For less extreme moral dilemmas, researchers found the responses of the group of six were much the same as the other volunteers.

below: An MRI scan reveals the medial prefrontal cortex [image courtesy of Dr Monica K. Hurdal, Department of Mathematics, Florida State University]

Too much of a good thing

Women looking for a partner regard attractive men with high-status jobs and a high income as "too good to be true". According to a study, given the choice between a handsome company director and a handsome travel agent, a woman will opt for the man who books flights in preference to the boss who takes them. While women welcome the handsome company director for his good looks, he is marked down for having a highly successful career.

Psychological tests suggested that when judging a man's suitability as a partner, women unconsciously take into account how likely it is that he will stray. Dr Simon Chu of the University of Central Lancashire in the UK said it seemed that women felt the company boss more likely to be unfaithful than the travel agent. He was similarly suspected of being too devoted to his demanding job, with all the extra hours involved, to spend much time at home.

Given the choice, a woman will opt for the man who books flights in preference to the boss who takes them

Dr Chu and researchers from the University of Liverpool assessed women's preferences by asking 186 female students with an average age of 23 to judge the desirability of men in lonely hearts adverts. The newspaper advertisements were put together by the research team to represent a cross section of attractive, average and unattractive men in jobs with high, medium and low status. Company directors, architects and lawyers were among the six high-status jobs, while travel agents, social workers and teachers were regarded as medium. Low-status jobs included waiters, gardeners and postmen.

The attractiveness of men whose photographs were used in the research had been assessed earlier in a pilot study. Mug shots of men were displayed on a screen alongside the words of randomly chosen lonely hearts adverts, and the women volunteers were asked to rate each of them in terms of appeal as a long-term partner on a scale of one to nine. Good looks guaranteed high scores, but Adonises with medium-status jobs came out as the women's overall first choice, and handsome waiters were ranked as just as desirable as the physically attractive company bosses. The researchers concluded that for all their advantages, men who combine wealth, power and still manage to be jaw-droppingly handsome are regarded as just too much of a good thing.

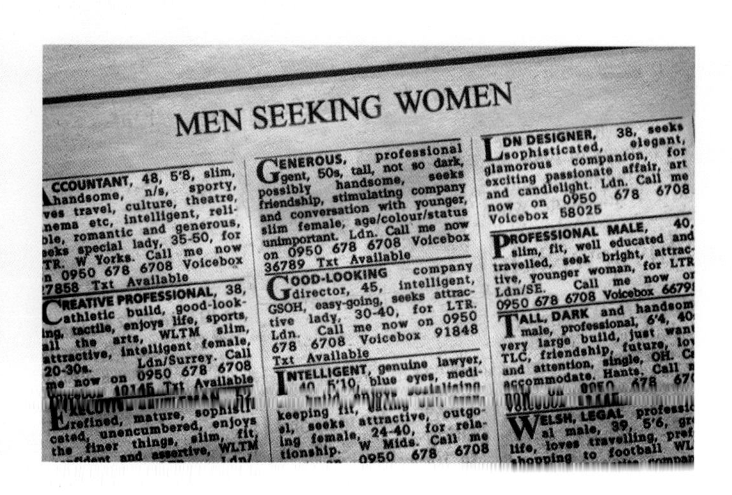

MEN SEEKING WOMEN

ACCOUNTANT, 48, 5'8, slim,
handsome, n/s, sporty,
ves travel, culture, theatre,
nema etc, intelligent, reli-
ble, romantic and generous,
eeks special lady, 35-50, for
TR. W Yorks. Call me now
n 0950 678 6708 Voicebox
7858 Txt Available

CREATIVE PROFESSIONAL, 38,
athletic build, good-looking, tactile, enjoys life, sports,
all the arts, WLTM slim,
attractive, intelligent female,
20-30s. Ldn/Surrey. Call
me now on 0950 678 6708

EXECUTIVE ... sophisti-
refined, mature, enjoys
cated, unencumbered, slim, fit,
the finer things, WLTM
and assertive, Ldn/

GENEROUS, professional
gent, 50s, tall, not so dark,
possibly handsome, seeks
friendship, stimulating company
and conversation with younger,
slim female, age/colour/status
unimportant. Ldn. Call me now
on 0950 678 6708 Voicebox
36789 Txt Available

GOOD-LOOKING company
director, 45, intelligent,
GSOH, easy-going, seeks attractive lady, 30-40, for LTR.
Ldn. Call me now on 0950
678 6708 Voicebox 91848
Txt Available

INTELLIGENT, genuine lawyer,
40, 5'10, blue eyes, medi-
keeping fit, outgo-
el, seeks attractive, for rela-
ing female, 24-40, Call me
tionship. W Mids. 0950 678 6708

LDN DESIGNER, 38, seeks
sophisticated, elegant,
glamorous companion, for
exciting passionate affair, art
and candlelight. Ldn. Call me
now on 0950 678 6708
Voicebox 58025

PROFESSIONAL MALE, 40,
slim, fit, well educated and
travelled, seek bright, attractive, younger woman, for LTR
Ldn/SE. Call me now on
0950 678 6708 Voicebox

TALL, DARK and handsome
male, professional, 6'4, 40
very large build, just want
TLC, friendship, future,
and attention, single, OH.
accommodate. Hants. Call

WELSH, LEGAL professional
al male, 39, 5'6,
life, loves travelling,
shopping to football

opposite: George Clooney as Dr Doug Ross on the television series ER—handsome and successful, but is he really what a woman wants? [photo: Getty / Handout]

right: Women browsing lonely hearts adverts showed a tendency to stray from those who described themselves as highly successful, instead favouring men who held middle-status jobs [image courtesy of Hayley Williams]

Pavlov's cockroaches

Cockroaches have been trained to drool at the prospect of food, just as Pavlov's dogs once did. Instead of the bells used for dogs, aromas were used to condition the cockroaches—but the results were much the same. After being taught to associate the smells of peppermint and vanilla with a forthcoming meal of sucrose, the cockroaches would dribble in anticipation.

Dogs, famously trained by Ivan Pavlov, and humans have previously been conditioned in experiments to salivate, but it is the first time such a response has been demonstrated in insects. By illuminating how the cockroach brain works, including the neural mechanisms by which salivation can be conditioned, scientists hope to gain insights into how the human mind functions.

Dr Makoto Mizunami, of Tohoku University in Japan, led the experiment and found that the cockroaches' conditioned memory lasted for at least a day. Cockroaches of the *Periplaneta americana* species were selected for the tests, with the individuals chosen on the basis that they failed to display any salivatory response when first exposed to the aromas of peppermint or vanilla. By wiping sucrose on to the insects' mouthparts whenever they were exposed to the smells, the researchers managed to teach the cockroaches to associate the two aromas with food. Once the conditioning was carried out, the researchers measured the quantity of saliva produced and found it increased noticeably whenever a peppermint or vanilla aroma wafted over the cockroaches.

Dr Mizunami said that while Pavlov's experiments took place more than a century ago, all the processes involved in the brain have yet to be understood. By looking at the cockroach's brain, which is much simpler than a mammalian brain, he hopes to identify what is happening in the mind when learning is taking place.

Cockroaches were taught to dribble in anticipation of a forthcoming meal of sucrose

opposite top: cockroach
[photo courtesy Steve Williams]
opposite bottom: False-colour scanning electron micrograph of the American cockroach, *Periplaneta americana*
[image: David Scharf / Science Photo Library]

Orderly minds

Blind people have superior memories to the sighted and a particular skill in recalling sequences, tests have shown. Experiments conducted at the Hebrew University in Jerusalem, Israel, resulted in sighted volunteers being trounced in comparison to the performance of blind people. Volunteers with and without the ability to see were asked by researchers to recall lists of 20 words, all of them Hebrew nouns, which were read aloud to them. Some of the tests simply asked for as many words as possible to be remembered, while others demanded that the order in which the words were read out also be recalled. Results showed that while the blind showed a small but clear superiority in the simpler tests, when it came to recalling sequences they were streets ahead of the sighted volunteers.

The researchers concluded their finding meant that the power of recall was helping the blind compensate for their lack of vision. It was not an innate superiority, but a learned skill that demonstrated the truth of the adage that practice makes perfect.

Because the blind cannot see, they have to rely on other means to navigate along a street or around a room. As objects cannot be recognised by looking at them, the blind remember them in sequences. A can of beans and a can of tomatoes are easily distinguishable by sighted people because all they have to do is glance at the label.

Reading the label is not an option for the blind, so they devise patterns that can be remembered, such as putting beans in the third row in the left-hand cupboard and the tomatoes in the eighth row.

Ehud Zohary of the Hebrew University said the blind view the world as a sequence of events in which objects are remembered in relation to each other. The same technique is useful in remembering long lists, hence their supremacy in recalling the 20 words.

The blind view the world as a sequence of events in which objects are remembered in relation to each other

below: Sheet of braille
[image: © iStockPhoto / Roman Milert]

Never again

It takes just one-tenth of a second for the brain to respond to visual clues that indicate we are about to repeat a mistake

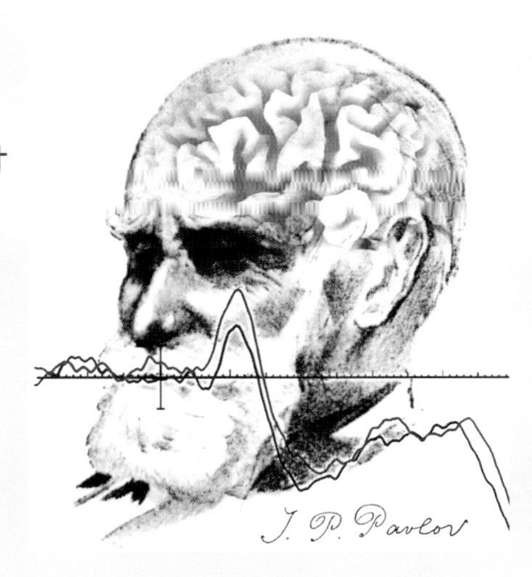

right: Ivan Pavlov and brain recordings from the study below, from top: The sequence of events participants were shown in Andy Wills' research; data showing the level and estimated source of recorded brain activity [images courtesy of Andy Wills, University of Exeter]

A tenth of a second is all it takes for our brains to register that we are in a situation where we have previously made a mistake. Measurements of mental activity have revealed that one area of the brain that recognises the similar circumstances is the lower temporal region, which lies by the temples. It identifies factors that were most likely to have caused the previous mistake and works, in effect, as an alarm bell that helps prevent a repeat of the error. It takes 0.1 seconds for the region to respond to the visual clues, providing an early warning that kicks in before there is time for conscious consideration.

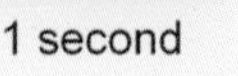

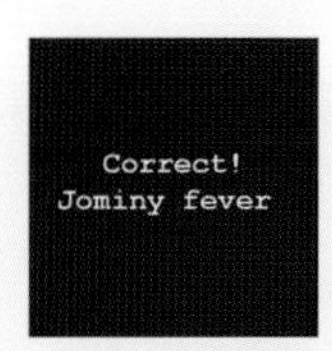

Psychologists at the University of Exeter in the UK made the discovery by placing electrodes on the scalps of volunteers. The electrodes detected changes in mental activity as the volunteers were asked to make predictions based on limited information, helping researchers work out both the speed and the location of the mechanism. Many of the volunteers' predictions were wrong, but when they were asked to make the predictions again after being told of the errors made the first time they carried out the exercise, activity was registered in the lower temporal region. By recognising the similarities of the current situation to the previous one, the volunteers were provided with a mental shortcut, helping them avoid the error they had committed earlier.

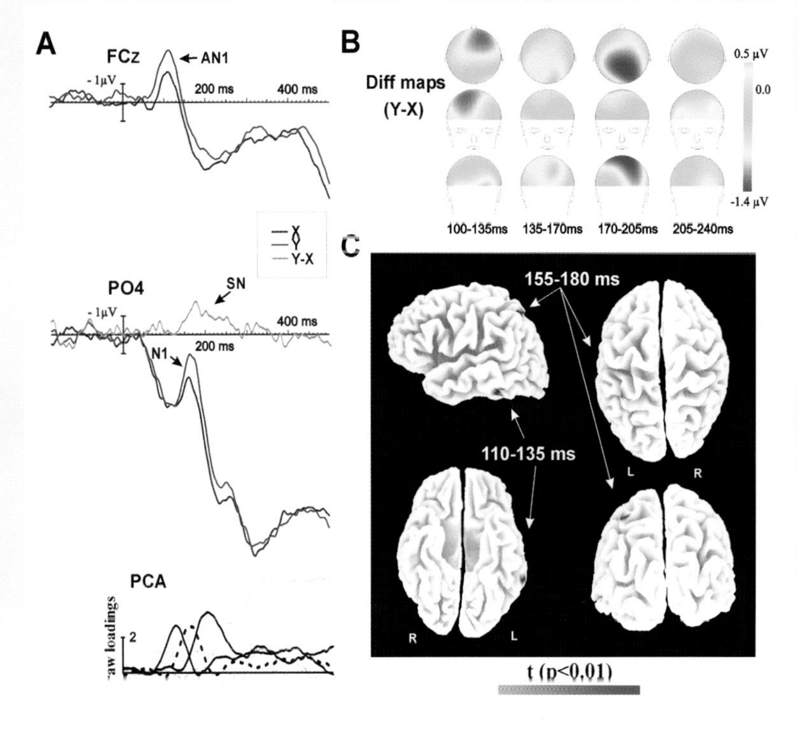

Professor Andy Wills, of the University of Exeter, led the research and said it was the first study to demonstrate how "amazingly rapid" the brain was in highlighting the likely causes of a previous error.

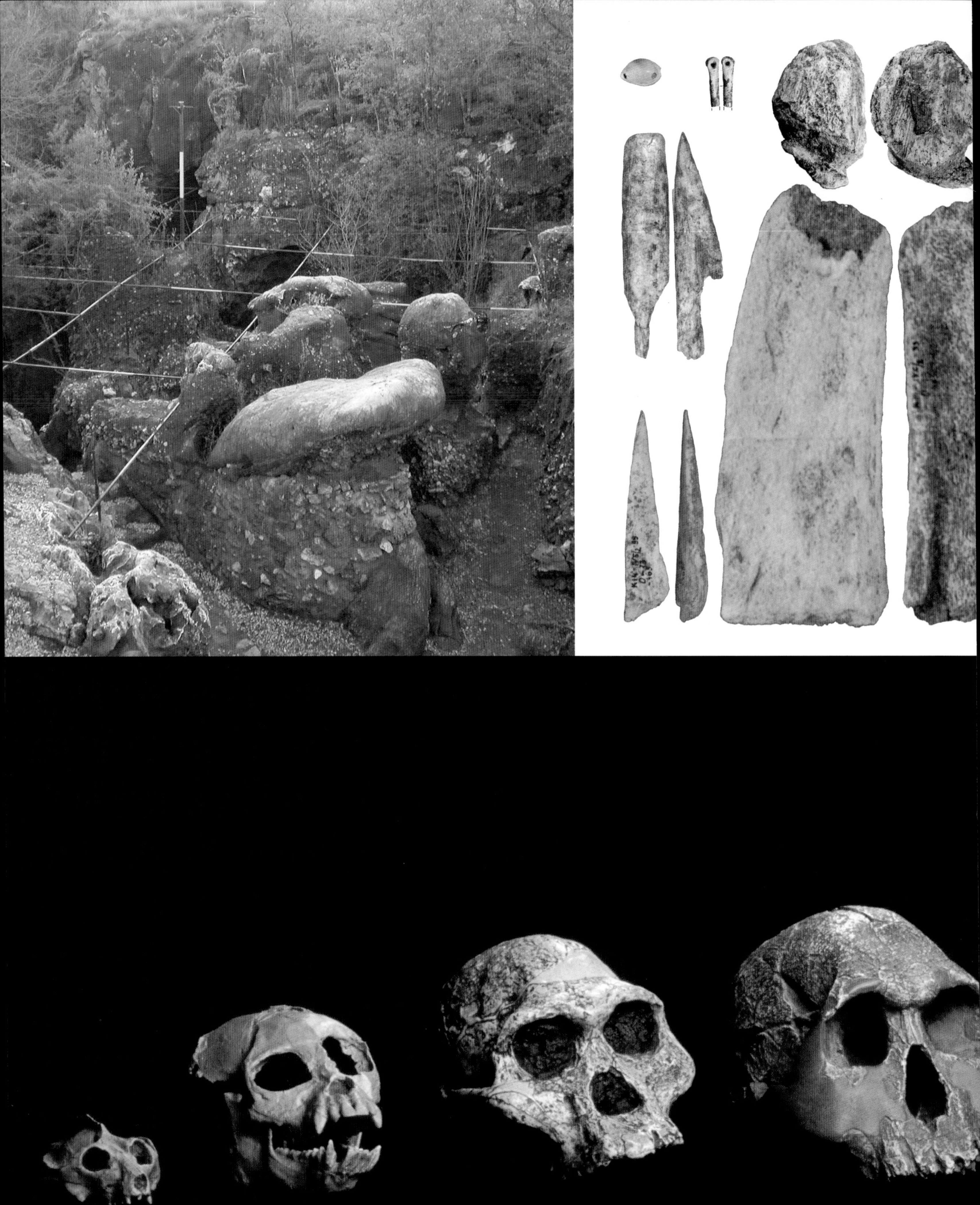

previous, clockwise from top left: Archaeological site in South Africa where remains of *Paranthropus robustus* were found [photo courtesy of Matt Sponheimer], artifacts from Kostenki [photo courtesy of John Frank Hoffecker], chimpanzee gesturing [photo courtesy of Frans de Waal / Living Links Center], seven skulls of ancestors and relatives of modern humans [photo: Pascal Goetgheluck / Science Photo Library]

Ancient People

The study of man's ancient ancestors and distant relations is a fascinating and often controversial field. By reaching back through the ages, it is possible to get an idea of the people who preceded us, how they lived and what they looked like. Examination of fossil remains and the earth they are dug from provides a wealth of information about early man's life and environment, revealing patterns of behaviour that can be startlingly familiar and at other times utterly alien.

Looking back over a matter of thousands of years, scientists are building up a clearer picture of how man left Africa as *Homo sapiens* to colonise the rest of the world, and how he turned from a nomad into a farmer. Further back, over hundreds of thousands and millions of years, it is possible to conjure an idea of man's family tree and to suggest the point in his evolution at which he stopped being a hairy ape and became human, albeit a primitive version.

The field, however, is one that often arouses heated and occasionally bitter debate. Few discoveries, for example, have generated as much controversy as that of the so-called Hobbit, an apparently miniature human that may have coexisted with modern man until relatively recently. Much of the debate takes place because of how little we know about what happened in the long and distant past, but as new discoveries are made and fresh information is yielded, claims and counter-claims are strengthened or refuted and in turn fill in gaps in our knowledge

Last refuge

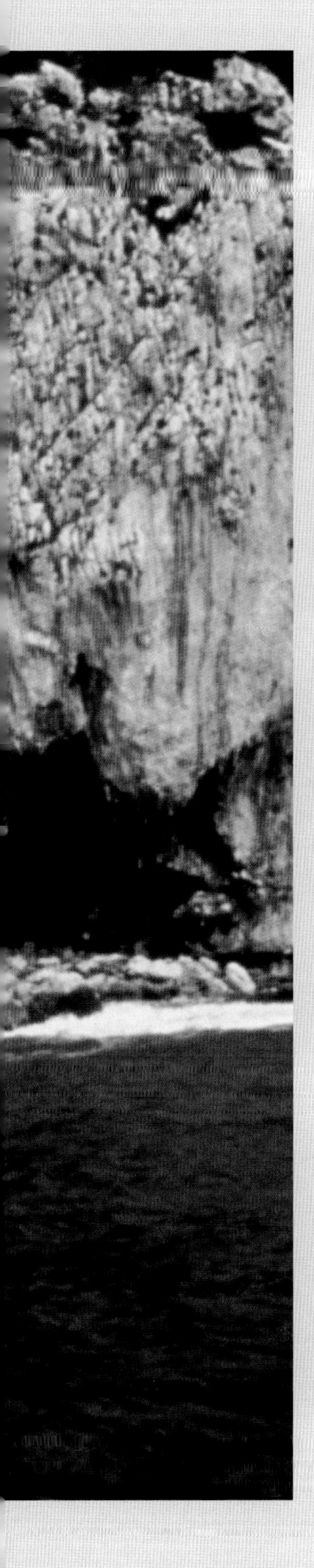

Gibraltar, the rocky outpost at the mouth of the Mediterranean, has been identified as the last known refuge of Neanderthal man. Neanderthal tools found in a cave at Gibraltar date to 28,000 years ago, and perhaps even as early as 24,000 years ago—the most recent date that Neanderthals have been confirmed as still surviving in a world being taken over by *Homo sapiens*. Until researchers dated the tools from Gorham's Cave, which include stone spear tips and knives, the most recent period the human relative, *Homo neanderthalis*, was known to have been alive was 30,000 to 33,000 years ago.

The finding adds weight to the theory that, rather than being the victims of widespread genocide committed by man's ancestors, Neanderthals went extinct from a mixture of causes. Genocide could have been one of the factors, though on a much smaller scale than has been suggested in the past. Climate change—with Europe cooling down as the ice sheets advanced during the last ice age, reaching their furthest extent about 20,000 years ago—is now thought to have been a significant, if not the main, cause of their demise. *Homo sapiens* are believed to have adapted better to cope with the climate changes when forest areas in parts of Southern Europe were transformed into tundra.

Bringing forward the date when Neanderthals were alive significantly extends the period they and *Homo sapiens*—who had reached Southern Spain at least by 32,000 years ago—coexisted in Europe. The ages were determined by radiocarbon dating the charcoal remains of the fires they burned in the cave. They range from 33,000 years ago to 24,000 years, but the earliest the research team from Britain, Gibraltar, Spain and Japan could be certain of was 28,000 years.

A mixture of tool types believed to have been used by the Neanderthals to cut up carcasses and scrape hides was found in layers of the charcoal more than eight feet (2.5 m) beneath the surface near the back of the cave, which spanned 131 feet (40 m). The cave was in many ways ideal for occupation by the Neanderthals, who used it for several thousands of years. It provided shelter from animals such as hyenas, was well ventilated with a high ceiling to stop the smoke from their fires [illegible], and daylight penetrated far inside. Evidence from the cave suggests that the Neanderthal occupiers cleaned it up repeatedly to keep it fit for habitation. Gorham's Cave is close to the sea level today, but when the Neanderthals occupied it 28,000 years ago it would have been almost 330 feet (100 m) above the water—giving them a spectacular view of the surrounding landscape of sand dunes, marshland, and the sea three miles (5 km) away.

Earlier finds reveal that the Neanderthals in Gibraltar would have eaten shellfish in the caves and the meat of creatures that roamed the region, perhaps including red deer, mountain goats, horses and rabbits. A meal identified in 1997 in another cave was thought to consist of pistachios, mussels and cooked tortoise.

Professor Clive Finlayson of the Gibraltar Museum was one of the lead researchers and said the Neanderthals at Gorham's Cave would have been protected from the worst of the climate cooling because of its location. He said the area would have been warm compared to most of the rest of Europe as the ice age intensified.

In a separate study, Professor Finlayson said the Neanderthals that took refuge at Gibraltar could have finally been driven into extinction by a sudden cold period 24,000 years ago, which tallied with the most recent charcoal radiocarbon date from the cave. The rapid chill 24,000 years ago was assessed as the most severe the region had experienced in 250,000 years. Oak and olive trees probably survived, but it was possibly the last straw for the less adaptable Neanderthals. *Homo neanderthalensis* and *Homo sapiens* are thought to have a common ancestor, *Homo heidelbergensis*, which they evolved from about 500,000 years ago.

left: Gorham's Cave, Gibraltar
[photo: Natural History Museum, London]

Cannibal

cousins

The combination of butchery marks and signs of starvation suggested that the Neanderthals were driven to eat each other to stay alive

Life for Neanderthals could be so tough that to get a square meal they would eat each other, butchery marks suggest. Analysis of eight skeletons found in Spain have revealed evidence they were hacked up for the meat on them—cut marks from stone tools on the bones closely resembled the pattern of cuts found on remains of animals butchered for their meat. Arm and leg bones had been smashed apart, apparently to get at the highly nutritious bone marrow, and some of the skulls appeared to have been deliberately cracked open to get at the brains. Examination of teeth showed hypoplasia lines, indicating that the individual Neanderthals went through several periods where they were on starvation diets.

The combination of butchery marks and signs of starvation suggested to researchers that the Neanderthals living in Spain 43,000 years ago were driven to eat each other to stay alive. The findings were made following the examination of bones of eight Neanderthals found in 2000 in a cave at El Sidrón in Asturias, Spain. Dr Antonio Rosas of the National Museum of Natural Sciences in Madrid led the research and said there was clear evidence of cannibalism.

What was less certain from the remains were the circumstances in which the bodies would have been cut up. He suggested it was most likely that the Neanderthals ate dead members of their group to improve their own chances of survival.

The hypoplasia lines in teeth from all eight individuals indicated that the Neanderthals were having trouble surviving in the region well before *Homo sapiens* are known to have reached the area as a potential competitor. A second possibility is that there was some ritual connected with the butchery, with the cut marks hinting at the spiritual life of Neanderthals.

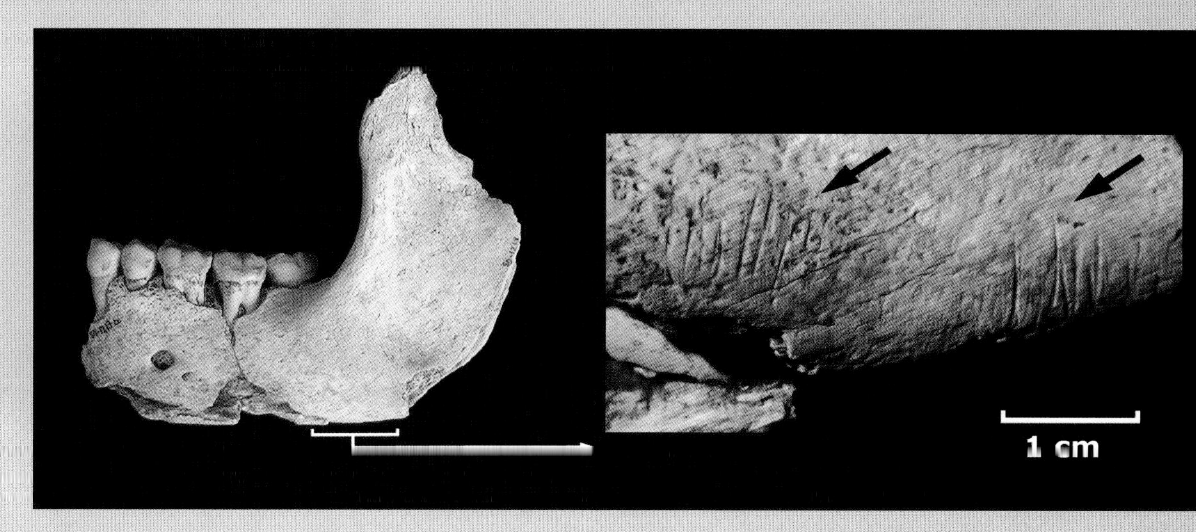

opposite: *Homo antecessor* practicing cannibalism [illustration: Mauricio Anton / Science Photo Library]
right: Specimen from El Sidrón, showing a close-up of cutmarks [photo courtesy of Antonio Rosas, Museo Nacional de Ciencias Naturales, CSIC]

Before the West was won

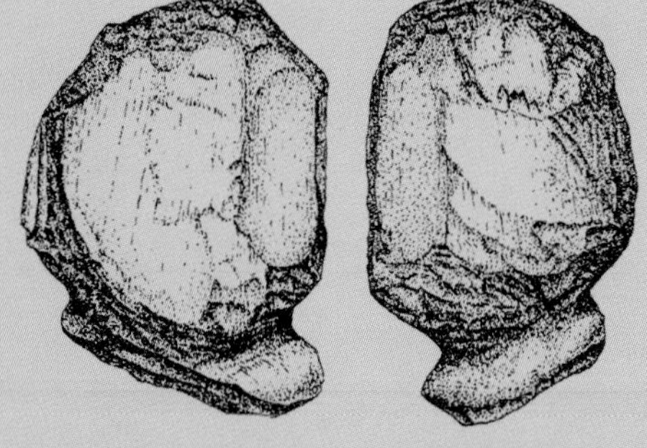

The ivory carving found in Russia has forced archaeologists to reconsider man's migration route from Africa

An ivory carving is helping reevaluate the date at which modern man reached Europe and the route he used to colonise the continent. Dated at up to 45,000 years old—the earliest record of modern man in Europe—the carving suggests that man moved northward into Russia before pushing westward towards the Atlantic during the migration from Africa. Archaeologists believe the figure represents an attempt to show a human head, making it the earliest known effort at figurative art, though they have accepted this theory as open to interpretation. It was carved from a piece of ivory taken from a mammoth, and the sculptor possibly discarded it before it was completed.

The ivory and other artifacts were found at a site in Russia called Kostenki, close to the River Don and 250 miles (402 km) south of Moscow. A layer of ash from a volcanic eruption 40,000 years ago in Italy covered the finds and helped archaeologists date them to 45,000 to 42,000 years old. The age of the discoveries astonished archaeologists because the location on the central East European Plain would have been cold, dry and inhospitable to the people who reached it on their migration from Africa. It was one of the last places evidence of human occupation would have been expected at that time, said Dr John Hoffecker of the University of Colorado at Boulder in the United States, who belonged to the international research team. Researchers, led by archaeologists from the Russian Academy of Sciences and the University of Colorado, said the finds were at least as early and quite likely earlier than traces of human occupation in Italy and Bulgaria, which meant the migration route from Africa needed to be reconsidered and perhaps completely redrawn.

Along with the ivory carving, which was broken in two, were two human teeth and primitive tools. The design of the tools indicated a technological leap forward, and the archaeologists said animal remains suggested the ancient people had learned to use nets for fishing and snares to trap Arctic foxes and hares. Reindeer and horse remains found at the site indicated that they, too, formed part of the diet.

Early forms of trading are thought to have been another feature of life on the plains because the stone used for toolmaking had to be transported from up to 100 miles (160 km) away. Similarly, there were shells used for jewellery that came from more than 300 miles (482 km) distant.

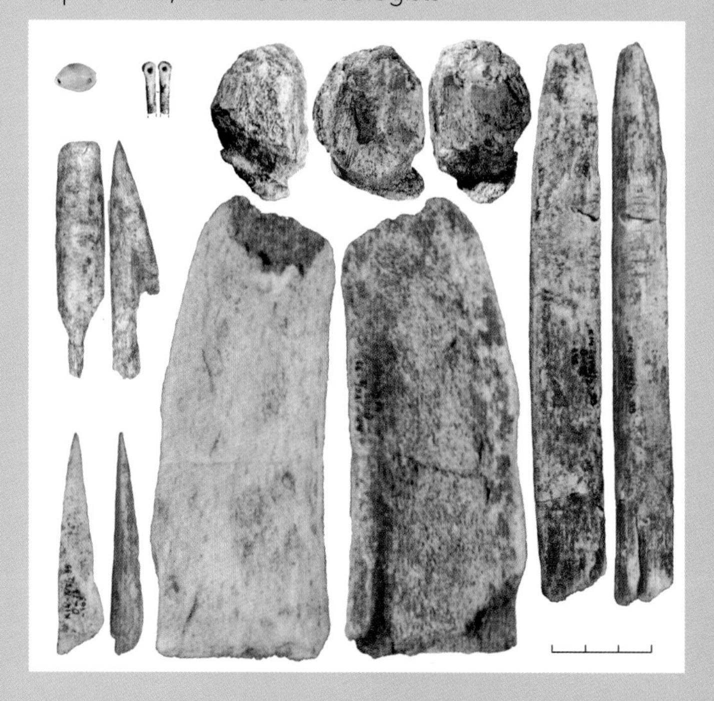

above: Sketch of carved ivory artifact from Kostenki, possibly the unfinished head of a figurine [courtesy of A. A. Sinitsyn]
left: Artifacts from Kostenki [photo courtesy of John Frank Hoffecker]

Weapons for the week

Female and young chimps used weapons to put them on an equal footing with the males

above: Tia, a female chimpanzee observed in the Iowa State University study [photo courtesy of Jill Pruetz]

Women rather than men would have been the first to fashion a weapon in early human groups, according to a study of chimpanzee behaviour. Observations of chimps in the Fongoli savannah in Senegal revealed that females and younger, weaker animals displayed the first evidence of systematic use of rudimentary spears among animals. They would break off a branch from a tree before trimming off any leaves or twigs and then stripping the bark away. The last stage was to sharpen one end with their teeth before jabbing the stick into holes in trees in hopes of spearing a bushbaby to eat.

The adult males were content to use their bare hands to catch and kill bushbabies, and it was only the weaker chimps that were seen to use sharpened sticks. The implication was that the females and young were using the weapons to put them on an equal footing with the males.

Dr Jill Pruetz of Iowa State University in the US led the study and said it had important implications for the evolution of the use of tools by people, and that it should at least lead to traditional explanations being reassessed.

Archaeologists have also unearthed evidence suggesting that chimpanzees, man's closest living relative, had a Stone Age just as humans did. Digs on the Ivory Coast exposed rocks still coated with the remains of the nuts for which they were used to smash open. The samples were retrieved from deposits that could be dated to 4,300 years ago—about the time the great sarsen stones were erected at Stonehenge in the UK—and were from an area that modern humans are not thought to have inhabited at the time. Moreover, the rocks were bigger and of a different shape to those typically associated with humans, while the nuts appeared to be of varieties eaten by chimps but not people today. The findings raise the question, but do not yet answer, whether toolmaking abilities were inherited by humans and chimps from a common ancestor, or if each developed it later.

Hobbit or red herring?

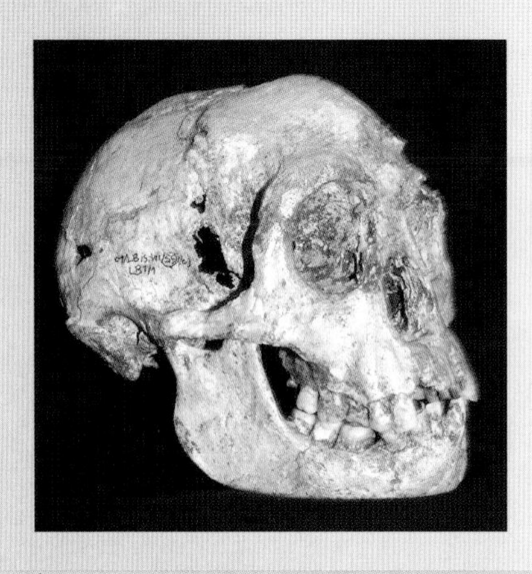

left: Half frontal view of the skull of *Homo floresiensis* [photo courtesy of Debbie Argue, Australian National University]
below: Illustration of human ancestors, *Homo erectus*. It has been suggested that *Homo floresiensis* descended from Homo erectus [illustration: Natural History Museum, London]

The sensational discovery of a hobbit-like creature in the Indonesian jungle has turned out to be one of the most controversial finds in decades. Some of the scientific interpretation of what the fossil bones represent has become so contradictory that the debate has come to echo the "oh yes he is, oh no he isn't" rows so beloved in British pantomimes. The original find was made by a joint Australian and Indonesian team on the island of Flores, east of Java. Since it was announced in 2004, it has been the subject of claim and counterclaim as scientists disagree about just what the fossils represent.

In the Liang Bua limestone cave at the western end of the island in September 2003, researchers unearthed the skull and fragments of bone of a small human-like creature, which came to be dubbed "Hobbit" after JRR Tolkien's popular and diminutive character, and officially called *Homo floresiensis*, a cousin of modern man. *H. floresiensis* stood upright but was only three feet (1 m) tall, and evidence showed the creature used tools, learned how to use fire, and was thought to have hunted dwarf elephants.

The discovery caused shockwaves in the scientific world because the Hobbit and other bone fragments from the site were calculated to have been on the island from 94,000 years ago to as recently as 12,000 years ago. The skull itself was dated as 18,000 years old. This meant that modern humans—who are thought to have reached the region between 35,000 and 55,000 years ago—could have lived alongside the hobbit-like creatures, who would have been the size of a three-year-old child. The earliest evidence of modern man on Flores itself, also found at Liang Bua, dates to 12,000 years ago. Before the finds were made by teams directed by Professor Radien Soejono of Indonesia's Centre for Archaeology and Professor Mike Morwood of the University of New England in Australia, it had been assumed that only *Homo neanderthalensis*, for which the most recent survival date is 24,000 years, had lived at the same time as *Homo sapiens*.

The size of the brain—about a third of the size of a modern human brain and equivalent to a grapefruit—caused astonishment because it was thought to be far too small for a species that had developed such a range of stone tools associated with the fossils. Peter Brown of the University of New England, who helped analyse the finds, suggested that *H. floresiensis* descended from human ancestors, probably *Homo erectus*, which had reached the island 840,000 years ago. The fact they were so small was attributed to evolutionary dwarfism, the process of being forced to grow smaller in order to survive as a species because of environmental confines, in this case an island.

Having reached the island, the species, held to be a member of the human family tree, lived alongside Komodo dragons and an extinct dog-sized rat. Cooked remains of stegodonts, primitive dwarf elephants approximately the same size as ponies, were found with the Hobbit's bones, suggesting the human-like creature hunted and ate them. Stegodonts would still have been a challenge for the tiny people to hunt, and most of the remains suggest they targeted the smaller, juvenile dwarf elephants. Other remains reveal they also ate tortoises, birds, bats, fish, frogs, rodents and snakes.

The original skull was so small that it was initially thought to be from a child, until it was realised the adult molars had developed. The skeleton was calculated to be a 55-pound (25 kg) female who died at about 30 years old. *H. floresiensis* had thicker eyebrow ridges, no chin, and longer arms in proportion to modern humans, suggesting it spent time in trees. It was also speculated that their size, together with the late dating of the bones, might explain why so many stories of "little people" living in the forests survive on the island even today.

Shortly after the announcement of the discovery, however, the findings were challenged by suggestions that, far from being a new species, the hobbit-like people were in reality merely modern humans with the disease microcephaly, which restricts brain and skull growth. The hand of the species supporters was strengthened when remains from at least nine more hobbit people, and more bones from the original specimen, were uncovered on Flores.

Professor Morwood said the additional finds, announced in 2005, helped build up a clearer picture of the Hobbit's fire-making and hunting abilities and provided more convincing evidence that *Homo floresiensis* represents a distinct species. There is, he said, evidence of at least 13 individuals, including a near-complete skeleton.

The doubters, however, came back with a vengeance in 2006 when a study concluded that the hobbit brains were too small to have been caused by evolutionary dwarfism. Brains can shrink, they said, but only moderately and not by as much as it appeared to have done with the Hobbit. Dr Robert Martin of the Field Museum in the US calculated that for the brain to have shrunk to the size of the Hobbit's, the body would have had to decreased to just a foot tall. He said the stone tools associated with the Hobbit finds had never previously been linked to *Homo erectus*, but that they did have features only ever seen among technology created by *Homo sapiens*. Furthermore, he argued that features of the Hobbit's anatomy were consistent with a modern human suffering from microcephaly. Unless significant new finds at the site are uncovered, and perhaps even if they are, the row over the Hobbit's status as a diseased human or a separate related species could rumble on for years.

On the menu

The menu suggests the species was nomadic and ranged across a variety of habitat types, including grasslands and woodlands, in order to find the plants

left: Illustration of *Paranthropus robustus* in a typical habitat [courtesy of Matt Sponheimer]

Chemical traces found in ancient teeth have given insight into what was on the menu for hominids 1.8 million years ago. The ancient species *Paranthropus robustus*, believed to be an offshoot of the lineage that led to modern man, was shown to have a wide-ranging diet that varied from month to month. Fruits, nuts, grasses, tree leaves, seeds, sedges, herbs, tubers and roots were all eaten by the australopithecine species. The menu suggests the species was nomadic, ranging across a variety of habitat types, including grass and woodlands, in order to find the plants.

The hominid was previously suspected of going extinct because it was unable to adapt to environmental changes that reduced the availability of the tough, low-quality vegetation it depended upon. But analysis of four *Paranthropus* teeth from Swartkrans, South Africa, revealed the species was much less specialised than once thought. This increases the probability that, rather than die out because of changing conditions, the species was killed off by man's direct ancestors.

Diet could be determined by identifying the carbon isotopes that were absorbed from food during the hominid's lifetime—each of the plants eaten produced a different isotope. The laser ablation technique used to examine the teeth was so precise that researchers could see the isotopes as they were laid down. Lasers sliced off tiny fragments of tooth, which were analysed once vaporised. The technique allowed them an astonishingly detailed view of the often dramatic changes in diet from year to year, or even month to month.

Researchers suggested that the changes in diet are likely to have been driven by variations in rainfall and the onset of droughts and heavy flooding. Professor Matt Sponheimer of the University of Colorado in the US said dietary changes over such short timescales 1.8 million years ago had never been seen in such detail. *Paranthropus* is believed to have shared an australopithecine ancestor, which lived about 2.5 million years ago, with a contemporary *Homo* species that eventually gave rise to *Homo sapiens*. It would have been about four feet (1.22 m) tall, had legs adapted to a bipedal lifestyle, and weighed less than 100 pounds (45 kg). Its brain was only slightly bigger than that of a chimpanzee.

Stone-Age dentistry

Despite the drill being primitive, the technique was surprisingly effective at getting rid of decay

Evidence of the earliest dentistry has been uncovered at a graveyard in Pakistan, and is thought to have been carried out by a beadmaker. Researchers investigating graves up to 9,000 years old in Mehrgarh discovered 11 teeth bearing the marks of Stone-Age drills. The findings put back the age of dentistry by several thousand years from the 5,000-year-old drilled Neolithic molar found at a site in Denmark.

The holes were made by flint-tipped drills attached to a bowstring, which enabled them to spin rapidly in a technique used at the time to make holes in beads. Nine different people, adults aged from 20 to over 40, underwent the treatment, and one of the patients was prepared—or perhaps given no choice—to put up with the pain for three separate teeth. Another of the patients had the treatment twice.

Scientists from the Université de Poitiers and the Musée Guimet, both in France, found that the drilled holes were up to one-seventh of an inch deep (3.5 mm) and were as little as four one-hundredths of an inch (1 mm) in diameter, suggesting considerable precision. Four of the teeth, all molars, showed evidence of tooth decay, and researchers were confident the procedures were carried out for their therapeutic benefit, though they could not be certain.

Despite the drill being primitive and the patient having to bear the inevitable pain without anaesthetic, the technique was surprisingly effective at getting rid of decay. There was no evidence of fillings to plug the holes to prevent further decay, but Professor Roberto Macchiarelli, a palaeoanthropologist at the Université de Poitiers who led the research, suspected a tar-like or vegetable substance must have been used. He was certain that the drilling took place when the owners—of whom there were at least four men and two women—were alive, because there was evidence of later wear from chewing. People living at the time gave their teeth hard wear because of the wheat and barley newly introduced to their diet. When the grain was ground down with stones, abrasive minerals ended up in the food.

Plenty of flint drill heads were found near the graveyard, as were bone, shells and turquoise beads. The dentistry took place around Mehrgarh for 1,500 years before mysteriously coming to a halt with the onset of the metal age.

below: Case examples from Mehrgarh
[photos courtesy of L. Bondioli and R. Macchiarelli]

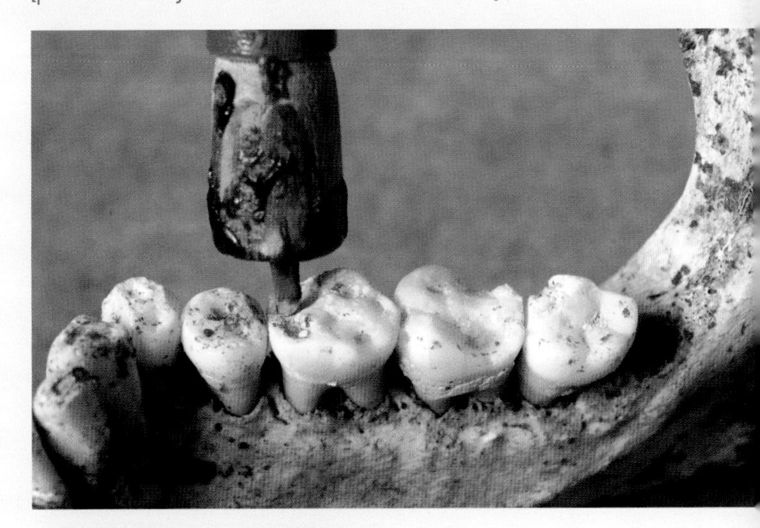

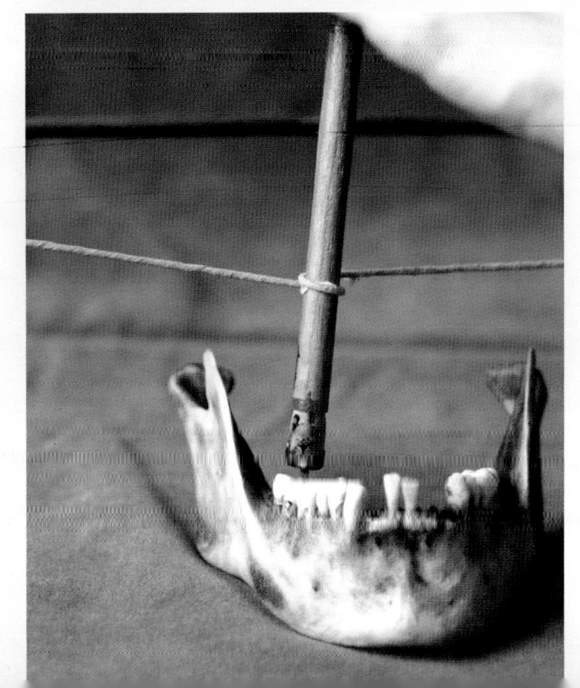

Observations identified 31 separate manual gestures, which is almost double the 18 vocal and facial expressions chimpanzees and bonobos use to communicate

above: A bonobo gesturing
right: Chimpanzees gesturing

Sign language

Hand and foot gestures are likely to have formed the earliest human language, according to a study of other primates. Chimpanzees and bonobos, the animals most closely related to modern humans, use gestures and demonstrate considerable control over their intended meaning. Observations of the animals have identified 31 separate manual gestures—almost double the 18 vocal and facial expressions they use to communicate. Some of the gestures are clearly understood even by humans, such as an outstretched hand to indicate a plea or a demand for an object to be handed over.

Gestures have the advantage of being more consciously controlled by primates, said Dr Amy Pollick and Dr Frans de Waal, of the Yerkes National Primate Research Centre at Emory University in the US, who carried out the study. Facial expressions and vocal sounds such as screaming are more closely linked to emotional reactions and more involuntary. Gestures are known to be a later evolutionary development than facial and vocal expressions, said the researchers, because they are present in apes but not in monkeys, from which apes separated from about 23 to 24 million years ago. They added that it is probable that a gesture used by chimpanzees, bonobos and humans alike would have been used by their common ancestor.

right: Chimpanzee gesturing [photo courtesy of Frans de Waal / Living Links Center]

Walking tall

To reach fruit in trees, orangutangs discard the four-legged walk and instead use two legs to walk along thin branches, grasping handfuls of twigs to balance

Man's ancestors had mastered the distinctive straight-backed, two-legged upright walk millions of years before they jumped out of the trees, a study suggests. The finding challenges the Darwinian notion that when early man first walked the ground, he started with a shambling, hunched, knuckle-walking gait and only over millions of years unbent into the vertical stance.

Scientists from the universities of Liverpool and Birmingham in the UK calculated the upright gait to have evolved some 17 to 24 million years ago. Their research challenges the previous idea that man's ancestors only started to learn to walk upright once they began to leave the trees for the forest floor four to eight million years ago. Researchers came to their conclusion after reassessing the fossil record and observing how the orangutan moves through the forest. They maintained that like the orangutan, man's ancestors would have benefited from walking upright in the trees while foraging for food.

Orangutans, as large apes, are too heavy to clamber on all fours along thin branches in the far reaches of trees where the majority of fruit is found. They can, however, reach much of the fruit if they discard the four-legged walk and instead use two legs to walk along thin branches while using their hands to hold on to twigs or vines to balance. The same technique is used to cross from tree to tree without the need to touch the ground where predators, such as tigers, might lurk.

In just this fashion, the scientists said, man's ancestors would have walked through the trees with the result that, when he came down to the ground, he could already stand with a vertical posture. They argue that this meant that man's ancestors would have avoided the need to ever have developed a knuckle-dragging walk, and that chimpanzees and gorillas have evolved it independently of other primates.

Professor Robin Crompton, of the University of Liverpool, and Dr Susannah Thorpe, of the University of Birmingham, led the study and said the gait used by chimpanzees and gorillas on the ground was an adaptation to allow them to wander about on the forest floor without surrendering their tree-climbing skills. The evolutionary move toward a vertical posture while still in the trees would have developed after apes separated from monkeys to take advantage of fruit instead of just relying on leaves, as monkeys then did.

Fossils, the researchers found, support their conclusion. They cited the *Morotopithecus* primate, which they said had developed the ability to walk on two legs 16 to 21 million years ago. They said man's ancestors probably started coming to the ground regularly as a result of climate changes that, from about 12 million years ago, caused forests to thin. The finding that bipedalism is not necessarily an indication that an extinct primate had left the trees and was a ground-dweller means that it may be harder to tell which fossils represent a human or the ancestor of a different type of ape.

opposite: A young male orangutan striking a very human pose in the jungles of North Borneo [photo: Co Rentmeester / Time & Life Pictures / Getty]

First fig

People recognised the potential of planting cuttings and thus began cultivating figs intentionally

below: Small figs from Iran,
bottom: Fig from Gilgal
[images courtesy of M. Kislev]

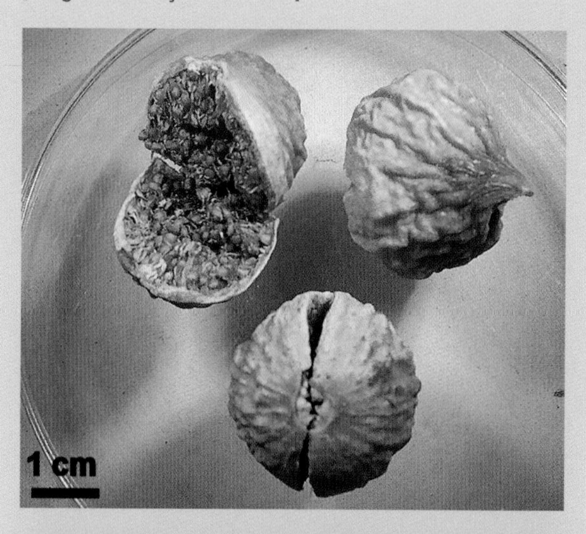

Figs rather than grain were the first crops harvested by man as he changed from a hunter-gatherer to a farmer, archaeologists have found. Carbonised figs dating back to 1,000 years before the first wheat and barley crops were planted were discovered at an ancient village in the Lower Jordan Valley close to Jericho in the disputed West Bank between Israel and Jordan.

Dated to as early as 11,400 years old at the village, Gilgal I, the figs were picked up to 5,000 years before it was previously believed they were domesticated. Hunter-gatherers would have picked and eaten fruit as they wandered through the landscape, but the figs at Gilgal I are believed to have been deliberately cultivated.

Archaeobotanists discovered that the nine small figs and 313 fragments of the fruit at the village were from a mutant variety. It was a parthenocarpic type that would produce ripe fruit without insect pollination and, as it was unable to produce seeds, it would die out with the tree. However, if cuttings were taken from the plant—as is thought to have happened at Gilgal—and put into the ground, the stem would sprout roots and grow into another fruit-producing tree. People recognised the potential of planting cuttings and thus began cultivating the fig intentionally, said Professor Ofer Bar-Yosef of Harvard University, who carried out the study with scientists from Bar-Ilan University in Israel.

The figs were found stored with other foods including wild barley, wild oats and acorns. Though the figs were carbonised, it appeared that they had been dried before being stored at Gilgal, which was occupied for 200 years from about 11,400 years ago.

Little Lucy

A three-year-old child who died, possibly swept away from her parents in a flood more than three million years ago, is providing unprecedented insights into human evolution. The remains of the female *Australopithecus afarensis* child were dug up in Ethiopia just two miles (3 km) from where Lucy, the most famous of the ancient hominids, was found in 1974. The youngster, dubbed Little Lucy, died about 150,000 years earlier than the adult Lucy, another of the *A. afarensis* species, whose nearly complete skeleton proved a watershed for the study of early humans.

Little Lucy dates from 3.3 million years ago and her discovery represents the earliest and most complete remains of a child from the human family. Previously only modern human and Neanderthal juvenile remains have been found in such a complete state. She was unearthed buried in rock that was formed from the sediments of an ancient flood that had either killed her or swept away her body soon after her death. She was rolled up in a ball and covered with sand before scavengers could get to her.

Analysis of her skeleton—much of which remains encased in rock that must be chipped away grain by grain—showed that she was able to walk upright, while the teeth revealed her age and gender. The thigh, shin and foot bones indicate her upright gait, but she had shoulder blades similar to a gorilla's and long, curved fingers for gripping, which suggests the species had retained the ability to climb trees. The combination of the upright walk on the ground and the tree-climbing ability indicates that the species was firmly on the path that led to modern humans. Researchers suggested that the species walked upright on the ground while foraging during the day, and used trees at night as a place to sleep and keep out of reach of predators.

While her face would have more closely resembled a chimp's than a human's, the brain is thought to represent a step towards modern man. CT scans revealed the size of her brain to be about the same as a three-year-old modern chimpanzee. Assessments of adult *A. afarensis* brains show, however, that Little Lucy's would have grown more than a three-year-old chimp's had she reached maturity.

Having the first complete juvenile skull of the hominid species has allowed researchers to determine how the brain changed from childhood to adulthood. Brain size was measured at 20.1 cubic inches (330 cu cm), which was 63 to 88 per cent of full size, whereas a chimp's brain is more than 90 per cent developed at the age of three. Scientists found that Little Lucy's brain growth rate was closer to that of modern humans than chimps, suggesting a possible behavioural shift for the species when they were living in the mixed savannah and forest environment more than three million years ago.

While the skull indicated an evolutionary shift towards the modern human brain, the voice box and the associated anatomy were still too primitive to have allowed Little Lucy to speak

Among the most exciting finds by the team led by Dr Zeresenay Alemseged of the Max Planck Institute in Leipzig, Germany, was the hyoid bone, which is fragile and has previously been found only in one specimen of a human ancestor, a Neanderthal. After analysing the shape of the hyoid, which supports the root of the tongue, the researchers concluded that, while the skull indicated an evolutionary shift towards the modern human brain, the voice box and the associated anatomy were still too primitive to have allowed Little Lucy to speak.

The first piece of Little Lucy was located in 2000 in the Rift Valley, south of the Awash River in the Dikika region of Ethiopia, when one member of the research team spotted part of the exposed skull. Her discovery and the initial findings were announced in 2006. Little Lucy is just one of the names she has been given. She has also been called Selam, a word meaning "peace" in several Ethiopian languages, and the Dikika Infant. It took four years to retrieve all the fossil bones and it is expected to take many more years of study before all Little Lucy's secrets are revealed and understood.

Hilltop calendar

Ancient Peruvians built the 13 towers of Chankillo long before the Inca civilisation rose to worship the sun

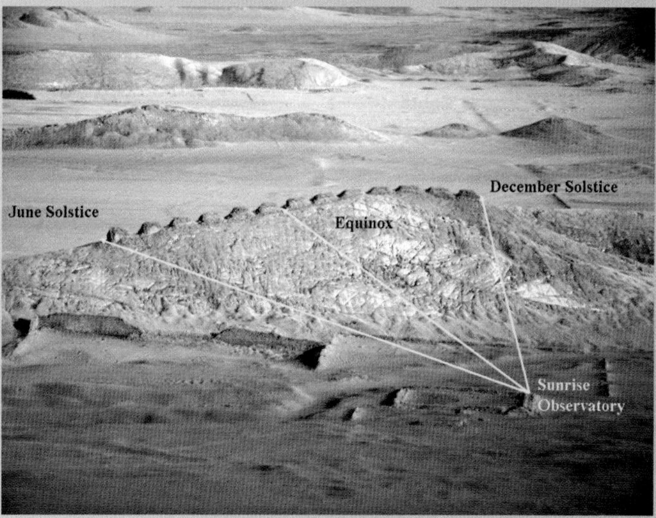

Thirteen towers lined up on the crest of a hill have been found to be a 2,300-year-old solar calendar. It is thought the towers were built by sun worshippers, and are positioned to show observers when the summer and winter solstices and the spring and autumn equinoxes take place. Ancient Peruvians built the 13 towers of Chankillo in the Andes long before the Inca civilisation rose to worship the sun.

The purpose of the towers—which range from heights of six to 20 feet (2 to 6 m) and extend 1,000 feet (300 m) along a hilltop—remained a puzzle until Dr Ivan Ghezzi, Archaeological Director of the National Institute for Culture in Peru, and Professor Clive Ruggles, an archaeoastronomer at the University in Leicester in the UK, carried out measurements. The solar observatory is the oldest in South America and hints at a civilisation prior to the Incas, where those who could claim to understand the sun were the ruling elite.

opposite: View of 13 towers of Chankillo
[photo © GeoEye / SIME / NASA]
right, top: The sunrise between Tower 1 and Cerro Mucho Malo at the summer solstice, 2003. The sunrise position at the solstice has shifted to the right by approximately 0.3 degrees since the year 300 BC
right, centre: Simplified diagram of how the solar observatory would have worked
[images courtesy of and © Ivan Ghezzi]
right, bottom: The fortified stone temple at Chankillo
[photo courtesy of National Aerophotographic Service, Peru]

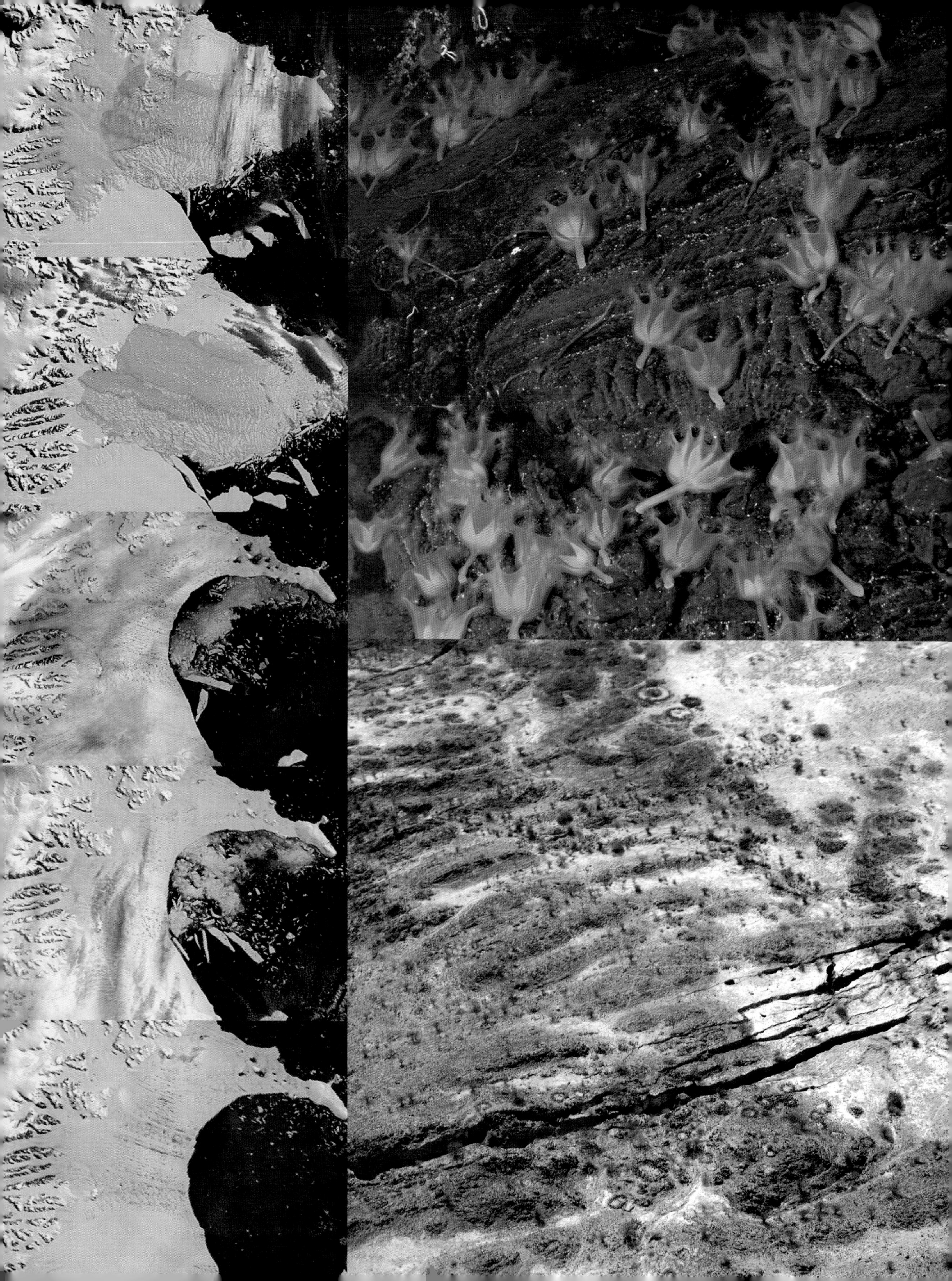

previous, clockwise from top left: Five images of the collapse of the Larsen B ice shelf of January, February, and March 2002 [courtesy of NASA / MODIS], pink, bell-shaped Stauromedusae jellyfish [photo courtesy of Emily Klein, Duke University], Researchers on the Beringia Expedition in 2005, which focused on ecological issues in the Arctic and was sponsored by the Swedish Polar Research Secretariat [photo courtesy of Martin Jakobsson, Stockholm University], logging in the Congo Basin [photo: NHPA / Martin Harvey], aerial photo of cracks and faults in the Afar Desert [photo courtesy of Julie Rowland, University of Auckland]

Shaping the Earth

Catastrophic hurricanes and flooding in North America, killer heatwaves across Europe, and the appalling death and destruction of the Boxing Day tsunami of 2004 in the Indian Ocean are all sobering reminders of the forces of nature. We may dominate the planet more than any other species in its history, but the awesome forces that have shaped the Earth are still occurring and we stand in their way at our peril. For the last 4.5 billion years since the Earth was formed from a cosmic soup, the planet has been frozen and baked, battered by asteroids and other space debris, and churned up over millennia by earthquakes and volcanoes on land and at sea. Continents have shifted, mountains have vanished, new oceans have formed and even the chemical content of the air has changed.

Man has doubtless had an influence in the short time he has walked the Earth, having built cities, flattened forests, constructed dams and formed an almost endless patchwork of fields across the landscape. With global warming, man has started to change the planet and its atmosphere in a way he never dreamed possible. But while climate change is the most immediate threat—and arguably the one we can do most to put right—there are others that could shape the world in even more dramatic and devastating ways, in which the Earth will survive but we might not. Huge meteors, super volcanoes, and natural climatic and atmospheric changes have all served to change the Earth in the past and will without question do so again.

A butterfly flaps...

The iceberg was smashed into half a dozen sections by the swell that had travelled 8,300 miles to reach it

A giant iceberg in the Antarctic was destroyed by a storm that took place almost half a world away in Alaska, researchers have found. For six days a swell generated off the coast of Alaska rolled southward across the Pacific Ocean until it reached the Antarctic and shattered an iceberg 60 miles (97 km) long. The iceberg, B15a, was smashed into half a dozen sections by the swell that had travelled 8,300 miles (13,300 km) to reach it. Previously, B15a had been half of the B15 iceberg, which became the biggest ever recorded when it broke away from the Ross Ice Shelf in Antarctica in March 2000.

It has been known since the 1960s that swells are capable of travelling halfway round the globe, but a swell's destructive effect on an iceberg had never been seen. When the B15 iceberg broke up on October 27, 2005, the weather was known to have been clear and calm, indicating that a storm in the immediate region was not the cause. US researchers, trying to follow the route of the swell that they knew had shaken apart B15a, were astonished when they discovered it had come from so far away. They were able to observe the breakup using satellite pictures and then flew to Antarctica to retrieve a seismometer, which had been left on the ice and recorded the frequency of waves and movements in the ice they caused.

Using data from the seismometer, the team of scientists, led by Professor Douglas MacAyeal of the University of Chicago and Professor Emile Okal of Northwestern University, were able to work out the source of the waves. It showed the storm-generated swell was powerful enough to lift the iceberg—which was 60 miles (97 km) long and 18 miles (29 km) wide—half an inch (1.3 cm) higher in the water and rock it four inches (10 cm) from side to side. Data from the seismometer revealed the iceberg was shaken for 12 hours before it shattered, and continued to pitch for another three days.

The immense degree of movement suggested a large storm, which the researchers set out to identify by calculating the wavelength distances and frequency of the swell. Wave-buoy records in Alaska and Hawaii tracked the waves en route to Antarctica, with further readings from a seismometer on Pitcairn Island in the South Pacific. In Alaska the waves were 35 feet (10.6 m) tall, and reduced to 15 feet (4.6 m) by the time they reached Hawaii two days later.

B15a had run aground near Cape Adare and the Possession Islands before being struck by the swell, and it is thought that it was perfectly positioned for destruction by the water. Professor MacAyeal described the point at which the swell became too much for the iceberg as "a fragile wine glass being sung to by a heavy soprano" .

Once scientists realised the potential of storm waves to reach Antarctica from so far afield, they started to look for other signs of the phenomenon. Among those identified was a typhoon in the Pacific. They also discovered that all 38 events big enough to register on the seismological meters from December 2004 to March 2005 were associated with distant storms in both the southern and northern hemispheres.

opposite: This aerial photograph of iceberg B15a was taken on November 24, 2005, just after the arrival of a sea swell from a large storm in the North Pacific generated six days earlier. The fragments seen in the photograph constitute approximately one-third of the iceberg's pre-fragmentation area [photo courtesy of Kathleen Lawson, Geophysical Institute, University of Alaska]
right: An infrared image of iceberg B15 from April 13, 2000, just one month after it broke off from the Ross Ice Shelf [image courtesy of the University of Wisconsin–Madison Space Science and Engineering Center, Antarctic Meteorological Research Center]

Vanishing

above: Logging in the Congo Basin continues, with more than 400 miles of roads built every year since 1990 [photo: NHPA / Martin Harvey]
opposite: Detail of logging, roads and forest cover in the Congo Basin [diagram courtesy of Nadine Laporte, Woods Hole Research Center]

forest

The effect of logging in the world's second-largest rainforest has been tracked by satellite and shows vast areas on the verge of catastrophic tree loss. The tropical forest in and around the Congo Basin in Africa is twice the size of France, but satellite pictures show tree clearance has quadrupled since 1990. For the first time researchers have been able to get accurate measurements of the logging expansion after assessing a series of satellite pictures taken from 1976 to 2003. They mapped 32,260 miles (51,920 km) of logging roads in the forest, which is spread over several central African nations including Cameroon, Central African Republic, Equatorial Guinea, Gabon, Republic of Congo, and Democratic Republic of Congo.

Dr Nadine Laporte, of the Woods Hole Research Centre in the US, led a team that calculated that the length of roads built through the forest by loggers rose from 97 miles (156 km) a year on average from 1976 to 1990 to more than 400 miles (640 km) a year since 1990. Trees are cut down up to half a mile (0.8 km) on either side of the roads, with hundreds of smaller tracks leading into the jungle. Observations show clearings in the forest caused by logging are six times more frequent than those that occur naturally.

In addition to the environmental changes brought about by chopping down trees, which is often done selectively rather than simply hacking down everything in sight, loggers bring even greater change by creating the roads that bring increased access to the area. Dr Laporte's team found that much of the damage to the forest has been comparatively limited because the majority of the logging had been taking place in sparsely populated areas. Logging roads are now, however, getting close to highly populated parts of the Democratic Republic of Congo, and this could have a disastrous affect on the forest because so many more people will have access to it. Wildlife will be the first to suffer, as bushmeat hunters will be able to penetrate far further into the jungle. The researchers calculated the area of forest now accessible to hunters is 220,000 square miles (570,000 sq km), or 29 per cent. It is feared that the next to come is agriculture, which will radically alter the landscape by removing trees wholesale to make way for fields.

One of the chief concerns about the loss of such forest cover is the impact it will have on the wider climate. Trees and other plants absorb carbon dioxide, the primary greenhouse gas that is held by the vast majority of the scientific community to be driving climate change. By absorbing carbon dioxide, trees help reduce the level of the gas in the atmosphere—but when they are destroyed they release it again. Deforestation is estimated to be responsible for 18 per cent of carbon dioxide emissions.

There are 32,260 miles of logging roads in the forest, which is spread over several central African nations

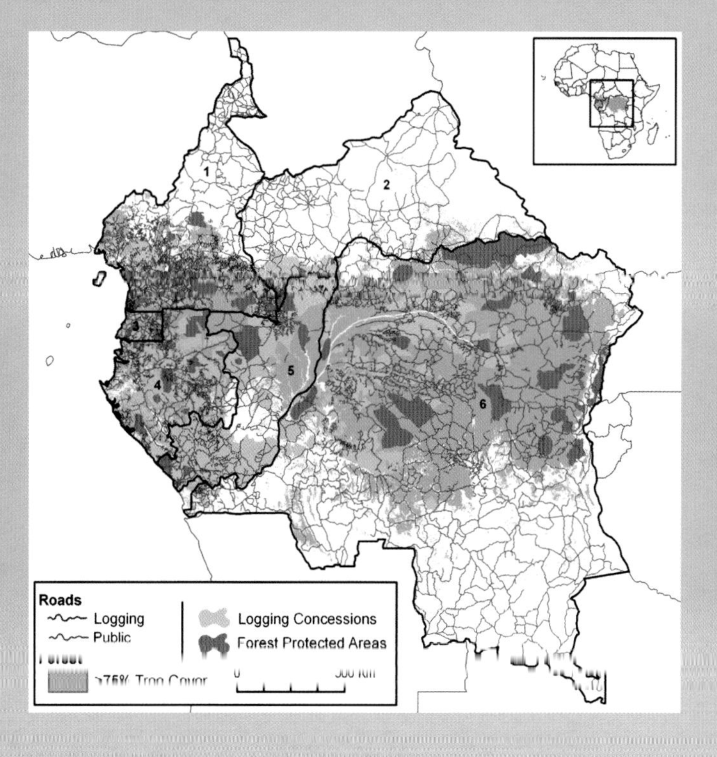

Deadly meteor

A meteor bigger and more deadly than the one implicated in the death of the dinosaurs was responsible for the greatest mass extinction on Earth, research suggests. Radar images from aircraft and gravity measurements taken by satellites have helped researchers pinpoint a crater 300 miles (480 km) wide under the Antarctic ice sheet. The crater is so big that only a meteor 30 miles (48 km) across could have created it, scientists calculated. By comparison, the meteor that produced the Chicxulub crater in Mexico and is blamed for ending the reign of the dinosaurs 65 million years ago is estimated to have been a mere six miles (10 km) across.

Located in the eastern part of the continent at Wilkes Land, the Antarctic crater was created by an impact some 250 million years ago, at about the same time as the Permian-Triassic mass extinction. During the extinction event, which eventually led to the evolution of primitive dinosaurs, an estimated 95 per cent of marine creatures and 70 per cent of land animals were driven to extinction. Professor Ralph von Frese, of Ohio State University in the US, said the impact of the meteor that created the Wilkes Land crater was so powerful that it strongly influenced the breakup of the Gondwana supercontinent. The impact was at a point where there is a rift valley extending into the Indian Ocean, and he suggested that when the meteor struck it prompted a chain of events that led to the separation of Australia from Antarctica.

The crater is so big that only a meteor 30 miles across could have created it

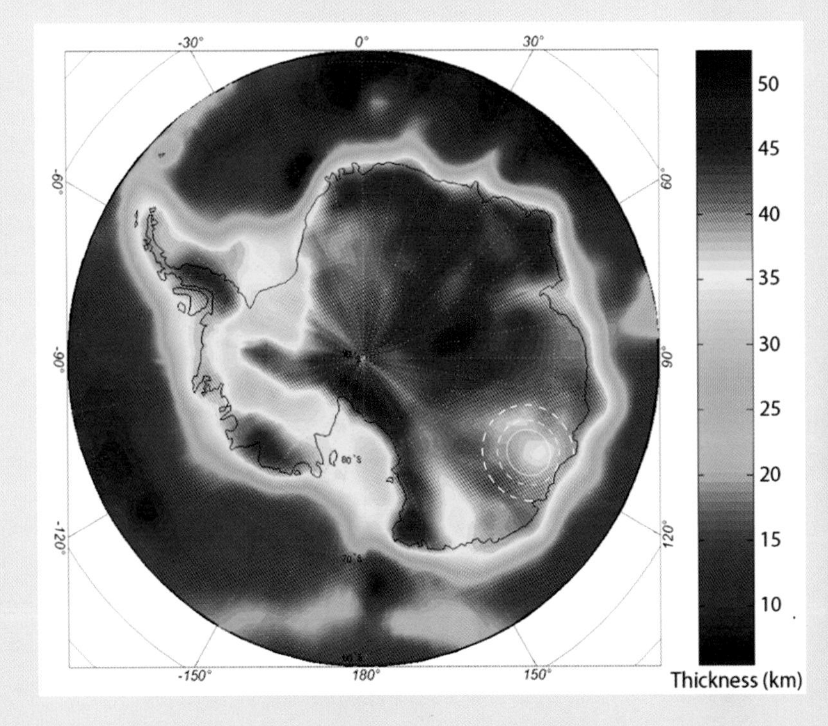

bottom left: The thickness of the Earth's crust across Antarctica, with thicker crust in red. The location of the Wilkes Land crater is circled (below, right of centre)
bottom right: Combined image of the Wilkes Land region of East Antarctica. The edges of the crater are coloured red and blue; a concentration of mantle material in the centre is colored orange
[images courtesy of Ohio State University]

Window on the Earth

A gaping and mysterious gash at the bottom of the Atlantic Ocean could prove to be a window to the secrets of the planet. Samples of rock from the seabed are being studied to solve the mystery of why hundreds of square miles of the Earth's crust are absent in part of the Atlantic.

Rock from the ocean floor was dredged and drilled by a team of scientists aboard a new state-of-the-art British research ship, RRS Cook, as part of the project to ascertain why such a huge section of the Earth's mantle is exposed. Researchers hope the rock samples and a host of observations from a robotic submersible will explain why the crust, which should form a seal over the mantle four miles (6 km) thick, is missing.

The absence of ocean crust over the mantle close to where the African, South American and North American plates meet defies conventional theories of plate tectonics. In general, as the Mid-Atlantic plates drift apart at a rate of less than an inch (about 2 cm) each year, the gap is filled with molten magma which erupts on to the seafloor and solidifies to become crust. In the vicinity of the Fifteen-Twenty Fracture Zone, however, bare mantle is exposed.

Dr Chris MacLeod of Cardiff University was one of the chief scientists on the voyage and described the exposed mantle as a "window into the Earth's interior". He said, "We hope to get a direct insight into the processes that go on in the Earth. It's an essential part of the basic understanding of how the Earth works."

By analysing the rocks and other data obtained from the Fifteen-Twenty Fracture Zone region, located between the Cape Verde Islands and the Caribbean, he and the rest of the team, including Professor Roger Searle of the University of Durham and Dr Bramley Murton of the National Oceanographic Centre Southampton in the UK, hope to gain insights into the mechanisms of how the underwater mountain chain of the Mid-Atlantic Ridge forms. In particular, they want to explain how the zone with the missing crust was created. The main suggestions are either that the mantle did not melt to form any crust when the plates pulled apart, or the crust was ripped aside.

As well as rock samples, the British research ship returned with detailed acoustic images of the ocean bottom where the Atlantic Ridge rises 13,123 feet (4,000 m) from the floor, which were taken by the submersible, the Towed Ocean Bottom Instrument (TOBI). Among the sensors was a side-scan sonar system to provide a map of the seabed that distinguished between areas of rock and sediment, and a magnetometer to date rocks and establish the speed of the movements of the Earth's plates by measuring variations in magnetic strength. A sub-bottom profiler allowed scientists to see through sediment to the rock below.

RRS James Cook, funded by the Natural Environment Research Council in the UK and the UK government, is arguably the most advanced research vessel in the world and was on its maiden scientific voyage.

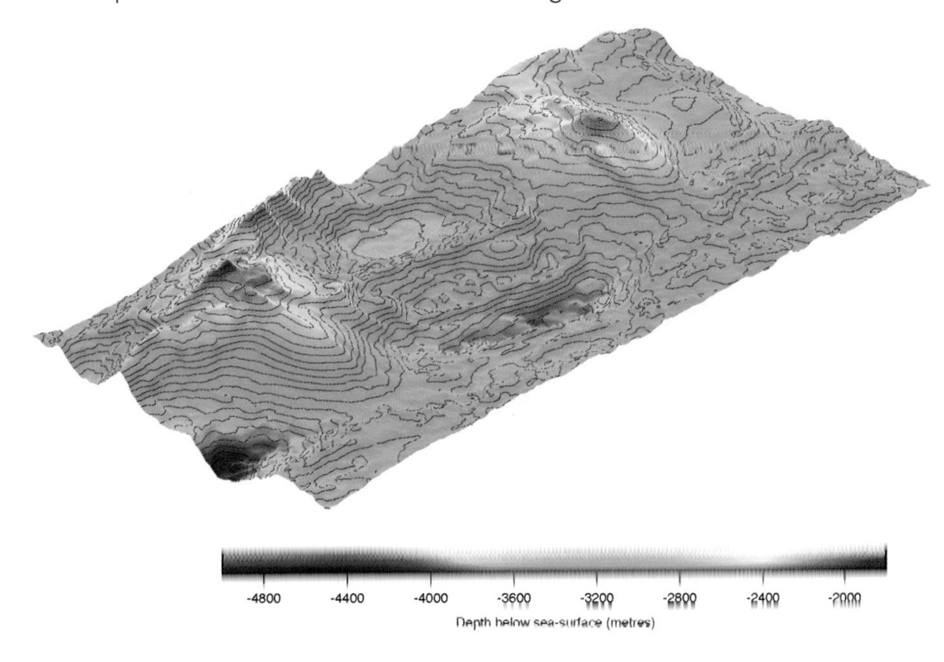

left: 3D perspective view of the seafloor elevation in part of the area in which the Atlantic ocean crust is missing. Dr MacLeod and colleagues found mantle rocks on the shallower turtle-back features (red to yellow/green) and believe they are probably being carried up on huge geological faults from far beneath the seafloor [courtesy of and © Roger Searle, Chris MacLeod & Bramley Murton]

Volcanic warming

Europe and North America were torn apart by long-lasting eruptions that took place along what is today Greenland, creating the North Atlantic

opposite: Two icebreakers and a drilling ship near the North Pole in late summer 2004 [photo courtesy of Martin Jakobsson, Stockholm University]

The most extreme global warming on Earth was caused by volcanic eruptions that lasted 10,000 years and gave birth to the North Atlantic Ocean, according to analysis of rock deposits. So dramatic was global warming 55 million years ago that up to half of the deep-sea animal species died out as the ocean surface warmed by up to 6°C (10.8°F), and the Arctic became a semi-tropical region where ferns thrived.

The source of the surge of greenhouse gas emissions that caused the warming, the Palaeocene-Eocene Thermal Maximum (PETM), has been identified by a study as huge quantities of organic matter lying in the path of volcanoes. Europe and North America were torn apart by the long-lasting eruptions that took place along what is today the eastern side of Greenland, creating the North Atlantic.

While the continents were pushed away from each other, volcanic magma boiled, bubbled and burned away huge natural stocks of dead plant and animal deposits, the study led by Dr Michael Storey of Roskilde University in Denmark concluded. As these vast compost heaps—many of them likely to have already been converted into coal, gas and oil fossil fuels—were cooked by the magma, they released the carbon trapped within them. Dr Storey calculated that over 10,000 years up to 4,500 gigatonnes of carbon was released back into the atmosphere, causing runaway global warming. The estimate ranges from 1,500 to 4,500 gigatonnes because of uncertainties about the form the carbon took. If it was derived, as suggested by previous theories, from methane hydrates (methane trapped by water in ocean sediments and the permafrost) becoming unstable and suddenly releasing large quantities of methane (CH4) gas, only 1,500 to 2,000 gigatonnes would have been required. A higher quantity is required if it was mainly in the form of carbon dioxide, which would have been released by sediments being cooked, as suggested by Dr Storey's theory.

The extra carbon in the atmosphere warmed the oceans by 5 to 6°C (9 to 10.8°F), and up to 8°C (14.4°F) in the Arctic where sea temperatures reached about 24°C (75°F). The researchers identified volcanic activity cooking organic material as the most likely cause of the PETM by dating a layer of volcanic ash in East Greenland and the Faeroe Islands that lies on top of basaltic lava in sequences up to four miles (6 km) thick. They were able to date the ash and lava by measuring the levels of trapped argon gas, which was formed over time by decaying potassium. The research team, from Oregon State and Rutgers universities in the US, observed that the rate at which greenhouse gases were pumped into the atmosphere during the PETM was slow compared to the rate today. While it took about 10,000 years during the PETM 55 million years ago for up to 4,500 gigatonnes to be emitted, the rate today is about eight gigatonnes every year and rising. Mankind's use of fossil fuels is widely held to be among the main causes of today's emissions, and at current rates it will take only about 600 years to match the worst nature could throw at the planet.

Sediment cores show the PETM warmed the seas up enough for the tropical algae *Apectodinium* to inhabit the waters surrounding the North Pole. Until the temperatures increased, the algae were restricted to lower latitudes. Dr Appy Sluijs of Utrecht University in the Netherlands was one of an international team studying 1,410 feet (430 m) of cores extracted from the Lomonosov Ridge in the Arctic between Siberia and Greenland as part of the Integrated Ocean Drilling Programme. He said that the cores showed temperatures in the Arctic waters rose from 18°C (64°F) to 23°C (73°F), more than 10°C (18°F) higher than earlier predicted for the region during the PETM, which is estimated to have lasted up to 200,000 years.

Analysis of the cores also revealed the presence of a freshwater fern, *Azolla*, 49 million years ago when the waters had cooled to about 12°C (54°F), suggesting that the Arctic basin was largely enclosed by land at the time and essentially cut off from the Atlantic. Professor Henk Brinkhuis, also from Utrecht University, led part of the sediment core research and said *Azolla* would have covered the water in dense mats, at least during the summer, and that the Arctic would have been a giant lake.

A further surprise thrown up by the cores, said Professor Kathryn Moran of the University of Rhode Island in the US, was the discovery of stones dropped from ice as early as 45 million years ago. It was previously believed that the Arctic had remained ice-free until about 15 million years ago, but the dropstones, released into the water by melting icebergs, suggested otherwise.

Creation of an ocean

The birth of an ocean has been witnessed as Africa is torn apart by the forces of nature. A huge rift has appeared in the Afar desert in Ethiopia as the African and Arabian continental plates pull apart and it will, according to geologists, one day be inundated by the Red Sea pouring in. Such processes are on average usually slower than the growth of fingernails, with plates moving a few fractions of an inch each year—but in September 2005 there was a spectacular shift.

Over a matter of a few weeks, hundreds of deep crevices appeared as the crust on either side was heaved away and the ground surged more than 26 feet (8 m) apart in places, almost overnight. It was calculated by a team of researchers from the University of Oxford and Royal Holloway, University of London in the UK, and Addis Ababa University in Ethiopia, that more than 2.6 billion cubic yards (2 billion cu m) of magma had poured into a crack and forced the plates apart. A series of violent earthquakes accompanied the separation of the continental plates, and 162 that registered four or more on the Richter scale were recorded in the space of a fortnight.

The separation of the plates is the same process that takes place in the Mid-Atlantic Ridge, but the sudden activity in the Afar desert has given geologists an unrivalled opportunity to see the process as it takes place. It was the first time a major rifting episode had occurred above sea level since the 1970s and the first time that before and after satellite images could be assessed to reveal the full extent of the changes.

Dr Tim Wright, now of the University of Leeds in the UK, and his colleagues researching the aftermath of volcanic and earth-shattering activity in the region concluded the birth of an ocean was being witnessed. As the rift widens and magma forces its way into the gaps, the future seabed is being created, though it will probably be a million years before it is part of a new ocean incorporating the Red Sea. Dr Wright said parts of Eritrea, Ethiopia and Djibouti are expected to sink to let in the sea, leaving parts of Northeast Ethiopia and Eritrea marooned as an island.

The future seabed is being created, though it will probably be a million years before it is part of a new ocean incorporating the Red Sea

above: 3D view of satellite radar measurements of how the ground moved in September 2005 [image courtesy of Tim Wright, University of Leeds, using Google Earth]
opposite: Image of the Dabbahu rift segment prior to the September 2005 events, enhanced to show subtle differences in rock type invisible to the naked eye [image courtesy of Ellen Wolfenden, Royal Holloway, University of London]

Photo of the eastern flank of the Dabbahu rift segment [Courtesy of Cindy Ebinger, Royal Holloway, University of London]

Chilled and boiled on the seafloor

Cold seeps and hydrothermal vents are among the most astonishing discoveries of the last 35 years. Not only were they extreme and unsuspected features of the ocean floor, but they were found to play host to a variety of marine animals and plants. The full range of animal and plant life that makes use of the features is still far from clear, and many of them are completely new. So little of the seabed has been visited by humans that it seems likely that plenty of vents and seeps, and perhaps new features, have yet to be discovered.

In the meantime, scientists who are willing to board deep-sea submersibles to look at the vents and seeps, and also those who send down robotic craft, are making a succession of finds. One of the more extraordinary discoveries was a shrimp of indeterminate species, though similar to the *Rimicaris exoculata* shrimp, which survived exposure to water heated to 80°C (176°F). Its preferred habitat appeared to be a narrow band of water heated to 60°C (140°F) by a thermal vent—but it was able to cope when even hotter water swirled into its path as it ate bacteria living on a hydrothermal vent. With recorded temperatures of up to 407°C (765°F), the vent was the hottest measured to date—warm enough to melt lead—and was located on the floor of the Atlantic Ocean's equatorial region. It was measured in 2006 by a research team led by Dr Andrea Koschinsky-Fritsche from the International University of Bremen in Germany.

Hydrothermal vents, first discovered in 1977, are features of the ocean floor that pump out seawater. The water is ejected in a hot stream, having already been heated by magma and absorbed chemicals. The ability of the shrimp and other creatures to live close to hydrothermal vents is a source of considerable interest among scientists who would like to find out why the proteins have not broken down in the heat. Temperatures within a few degrees of 407°C had been recorded previously, but the highest measurement came as a surprise. Paul Tyler, a professor of Deep Sea Biology at the National Oceanography Centre in Southampton, UK, said the vent's measurement challenged the limits of physics and chemistry.

Equally intriguing are cold seeps, which were first found in 1984 and leak out methane and sulphide, forming pools on the seafloor rather than mixing with the water. An expedition to the Antarctic led by Julian Gutt of the Alfred Wegener Institute for Polar and Marine Research in Germany as part of the Census of Marine Life, was able to take the first samples from a seep 2,625 feet (830 m) down, which was discovered in 2005 by a US survey team. Hundreds of shells of dead clams litter the seep, which is thought to have become inactive, though how long ago is a matter of considerable uncertainty.

With temperatures reaching 407°C, the vent was hot enough to melt lead

below: Undersea robot Jason II examines the smoking Medusa vent
opposite, top: Deep sea photos taken at the 407°C hot vent [photos courtesy of Andrea Koschinsky / MARUM / University of Bremen]
opposite, bottom: Unusual pink, bell-shaped Stauromedusae jellyfish thrive near the vent [photos courtesy of Emily Klein, Duke University]

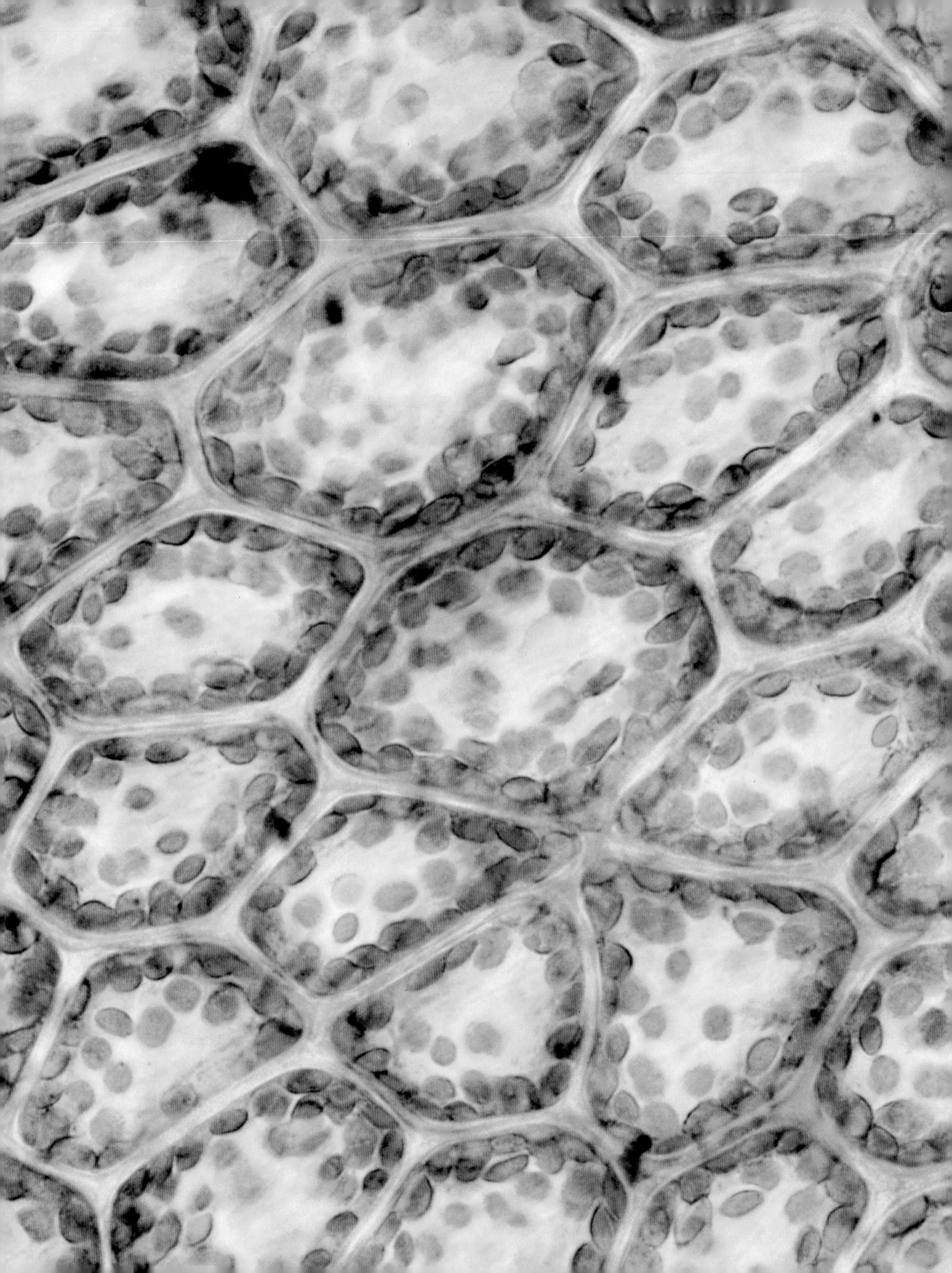

A world to breathe in

A mysterious lack of oxygen on Earth for 300 million years after the evolution of plant life has been explained. Photosynthesis, the process by which plants convert sunlight into food, pumps oxygen into the atmosphere as a waste product. Although plants began photosynthesising 2.7 billion years ago, the geological record shows the earth remained low in oxygen for another 300 million years. The "oxygen gap" between the advent of photosynthesis and the "Great Oxidation" mystified scientists because it seemed the fossil plant records and the chemical traces of the ancient atmosphere contradicted each other.

Researchers at the University of East Anglia in the UK have now established it was quite possible for the atmosphere to remain low in oxygen even in a world where plants were photosynthesising. They found that the production of oxygen through photosynthesis was not in itself a guarantee of a high oxygen content, but that another factor, such as a sudden release of oxygen from organic sediments, boosts the process. Once levels reached a certain point, ozone would have started to form and would have accelerated the accumulation of oxygen in the atmosphere, making it possible for more complex lifeforms to evolve.

Their finding has implications for the search for life in other solar systems and galaxies. Instead of considering only those planets high in oxygen and ozone, astronomers could investigate atmospheres with low levels and still have a chance of discovering lifeforms capable of photosynthesising.

The "oxygen gap" mystified scientists because it seemed the fossil plant records and the chemical traces of the ancient atmosphere contradicted each other

opposite: Light micrograph showing a section through a leaf, magnified 250 times
[image: John Durham / Science Photo Library]

Giant insects

Researchers found that the proportion of oxygen in the air limits the size of insects

The oxygen-rich atmosphere on Earth 300 million years ago made it possible for giant insects to flit, skitter and crawl around the ancient landscape. Extinct dragonflies with wing spans of 2.5 feet (76 cm) were among the creatures that could not exist today because of lower oxygen levels, according to research into modern beetles.

Oxygen forms 21 per cent of the air available to breathe today, but 300 million years ago it was 35 per cent. By studying different types of beetles, researchers found that the proportion of oxygen in the air limits the size of insects. They determined that as the insects get bigger, they have to devote a disproportionate part of their bodies to breathing. Instead of taking air down the trachea into the lungs for oxygen to be transported around the body by blood as mammals do, insects have a whole system of tracheae. Air is taken in through holes in their bodies, called spiracles, to the network of interconnecting tubes that deliver oxygen directly to all parts of the body.

Using X-rays to assess the inner workings of the insects, researchers found that the size of the tracheal system in the *Eleodes obscura* beetle, which is about 1.5 inches (3.5 cm) long, takes up 20 per cent more of the body than it does in the smaller species *Tibolium castaneum*, which is about one-tenth of an inch (2.5 mm) long. The tracheal network needed to become longer to reach the limbs, and the tubes grew in diameter because they had to take in a greater quantity of air since the demands for oxygen grew with the size of the beetle.

The research team, led by Professor Jon Harrison of Arizona State University in the US, found that there would be a critical point at which body size would have to stop expanding because the tracheal system would be too big to fit through the openings between the body and legs. Calculations by the researchers based on this critical point identified 5.9 inches (15 cm) as the maximum length of a beetle. This correlated closely with the size of the biggest known beetle living today, South America's Titanic longhorn beetle, *Titanus giganteus*, which reaches 5.9 to 6.7 inches (15 to 17cm).

Dr Alexander Kaiser, of Midwestern University and one of the lead researchers, said insects had the same body structure 300 million years ago, though beetles themselves were not around, and would have been able to grow narrower tubes because each breath would have taken in more oxygen than it does today.

opposite: *Titanus giganteus*, **the world's largest known beetle [photo: © Natural History Museum, London]**
below: X-ray contrast image of the darkling beetle ***Tenebrio molitor*** **[courtesy of Alex Kaiser, Jaco Klok, Wah-Keat Lee]**

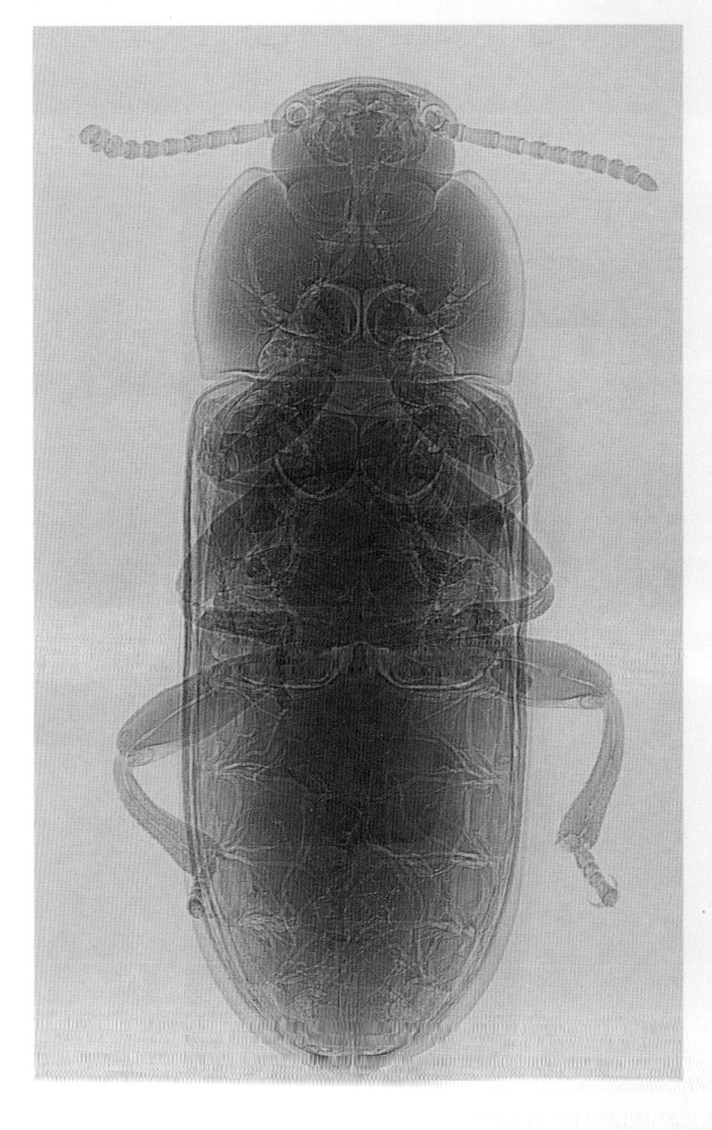

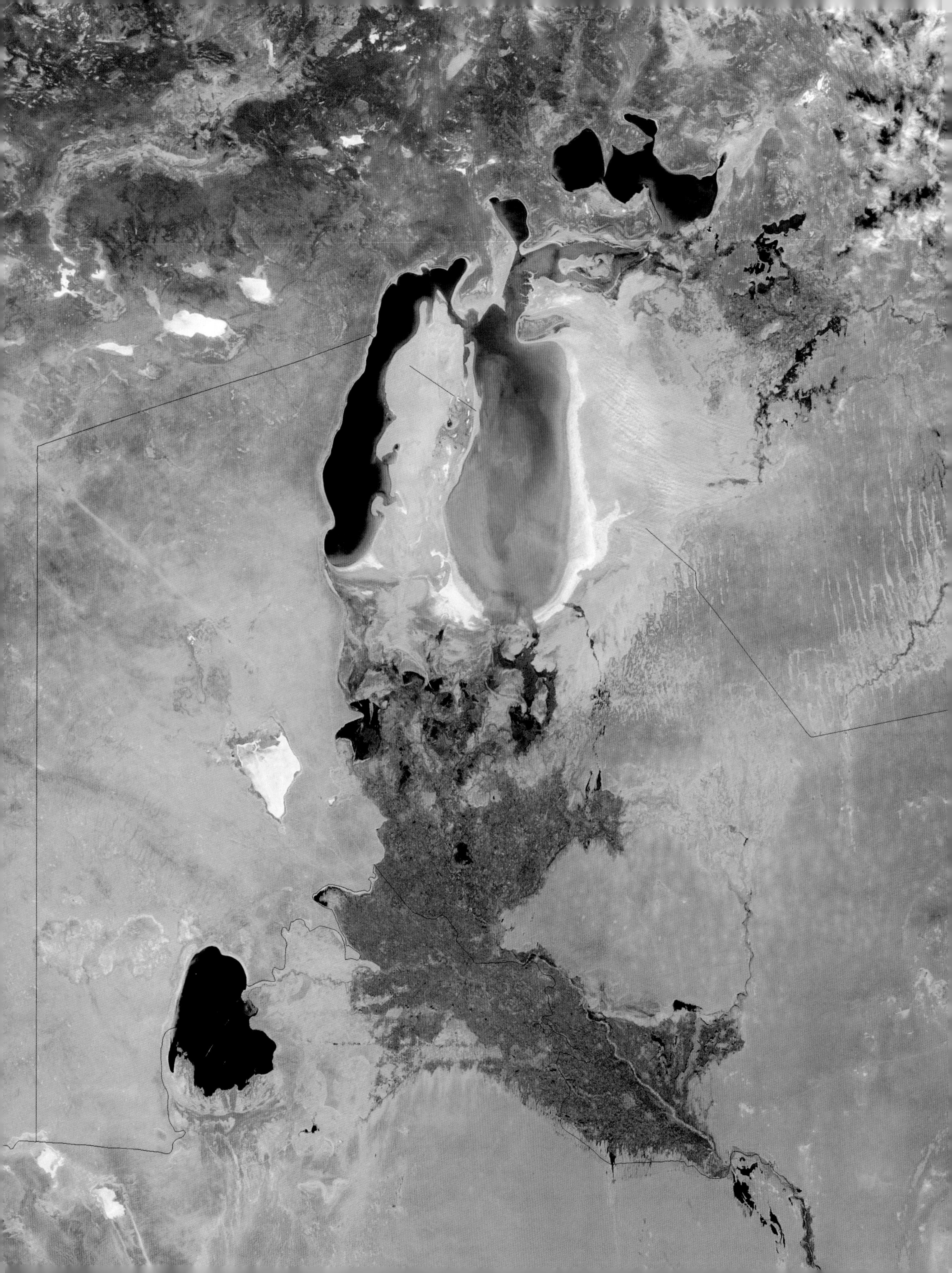

Resurrection of a dying sea

A sea that was driven to the brink of oblivion by engineers in the Soviet Union is being brought back to life. The Aral Sea, the world's fourth-largest inland body of water, was dismissed as a mistake of nature by Soviet engineers who wanted to redirect the waters flowing into it to sustain cotton farms. By diverting two rivers, the Syr Darya and the Amu Darya, they were able to irrigate valuable crops in desert areas of Kazakhstan, Uzbekistan and Turkmenistan—but the price was the loss of the Aral Sea.

From the 1960s to 2005 the Aral's waters dropped, leaving previously thriving ports miles from the water's edge and ships stranded in the desert. The Aral Sea was reduced to less than a quarter of the 25,500 square miles (66,000 sq km) it once covered and the water became too salty to support the freshwater fish that once provided a living to the people in surrounding towns and villages. Dust from the seabed, loaded with toxins from pollutants, swirled over settlements and is thought to have caused devastating increases in bronchial, arthritic and other diseases that cut the region's life expectancy from 64 to 51 years.

Waters are now rising again in the northern section of the Aral Sea and fish are returning after a dam was constructed to allow the Syr Darya to feed it once more. While the larger southern section is still drying out, the smaller northern section in Kazakhstan had by mid-2007 increased in size by about 385 square miles (1,000 sq km) to 1,250 square miles (3,250 sq km). Depths rose from 10 feet (3 m) to 140 feet (42.5 m).

Bringing the sea back to life has been achieved by building an eight mile (13 km) dyke at the Berg Strait and increasing the quantity of water that reaches it from the Syr Darya with remedial work to the banks. The government of Kazakhstan carried out the work with a loan from the World Bank, after two abortive attempts to build a dam that followed the country's independence from the Soviet Union in 1992. It is hoped that water will return to a wider area of the northern Aral Sea over the next few years. A second dam planned for the region, about 65 miles (100 km) north of the first, will be assisted by a canal cut from the Syr Darya in refilling the sea.

The Aral Sea was dismissed as a mistake of nature by Soviet engineers who wanted to redirect the waters flowing into it to sustain cotton farms

opposite: Image from NASA's Terra satellite of the Aral Sea, taken in October 2003 [courtesy of Jacques Descloitres, MODIS Rapid Response Team, NASA / GSFC]
below: These satellite images show the changes in the northern Aral Sea from April 9, 2006 (top) and April 8, 2005 (bottom) [courtesy of Jesse Allen, Earth Observatory, Goddard Earth Sciences DAAC]

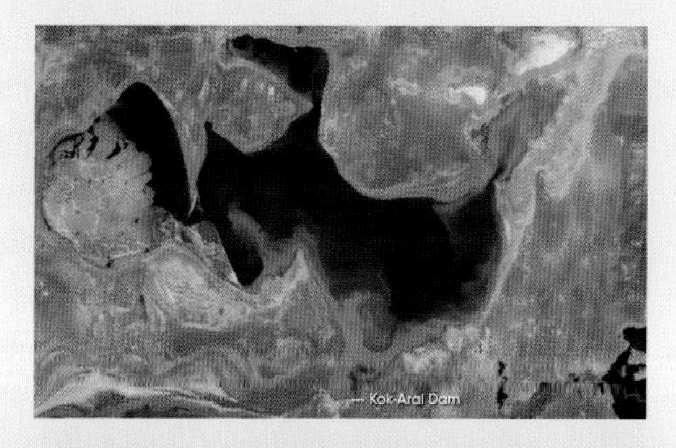

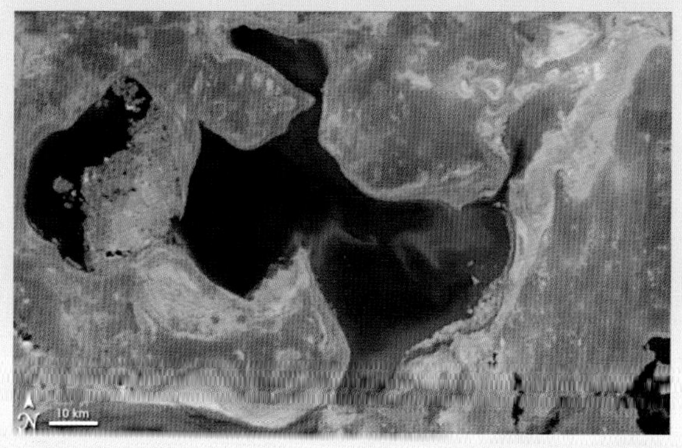

Arctic melting

Global warming is increasing temperatures all around the world but is most pronounced at the poles, which—though still cold by the rest of the planet's standards—are warming faster than anywhere else. In the business-as-usual scenario in which no attempt is made to cut greenhouse gas emissions, as assessed by the United Nation's Intergovernmental Panel on Climate Change (IPCC), it is forecast that global temperatures are likely to rise up to 6.4°C (11.5°F) on average by the beginning of the next century. At the poles, however, the increase would be even more pronounced and a temperature rise of 8°C (14.4°F) or more is quite possible.

In the Arctic, the extent of the ice in the summer has shrunk up to 9.8 per cent each decade since 1978. Average Arctic temperatures have already risen at twice the rate of the global average for the last century and temperatures in the top layer of permafrost have risen by up to 3°C (5.4°F) since the 1980s. The IPCC highlighted concerns that by the end of the century sea ice in the Arctic will virtually all have melted by the end of each summer.

A further study, funded by NASA and the National Science Foundation in the US, was completed too late to be assessed by the IPCC but found that the ice over the North Pole could be gone during the late summer weeks by as early as 2040. Some ice would remain on coastlines, such as those of Greenland and Ellesmere Island, but otherwise the Arctic sea would be open to shipping, the research team led by Dr Marika Holland of the National Centre for Atmospheric Research in the US concluded. A climate model simulation showed that the rate of ice loss is likely to remain steady until about 2024, when there will be a sudden increase in the speed of the process. By 2060 or earlier, the ice will in essence be gone in late summer. Professor Chris Rapley, head of the British Antarctic Survey, said the US study might even be understating the problem of retreating ice because emissions of carbon dioxide, the main greenhouse gas behind climate change, are still rising globally.

While the loss of sea ice would open up the Arctic Sea to shipping and the potential of oil exploration, the rising temperatures would have consequences for the human inhabitants of the region and the wildlife. Conservationists have highlighted the polar bear as likely to be one of the biggest losers of climate change, but all wildlife in Arctic regions would be affected.

The first study to measure climate change in terms of the arrival of spring in the Arctic found that it is now taking place more than two weeks earlier than it did just a decade ago. Researchers from Denmark looked at the flowering of plants, egg laying by birds and the appearance of insects at Zackenberg in Greenland to establish that spring is arriving earlier. In Northern Europe similar studies have shown that spring has advanced 2.5 days earlier per decade, while the global average is 5.1 days. Dr Toke Høye of the University of Aarhus in Denmark said the trend for the Arctic was surprisingly strong but confirmed suspicions of how the seasons are changing.

opposite: Signs of spring in Greenland, arriving earlier every year [photo courtesy of Toke Høye]

below: Minimum Arctic sea ice in 2000, left, and in 2040, right, as estimated by the Community Climate System Mode [courtesy of UCAR]

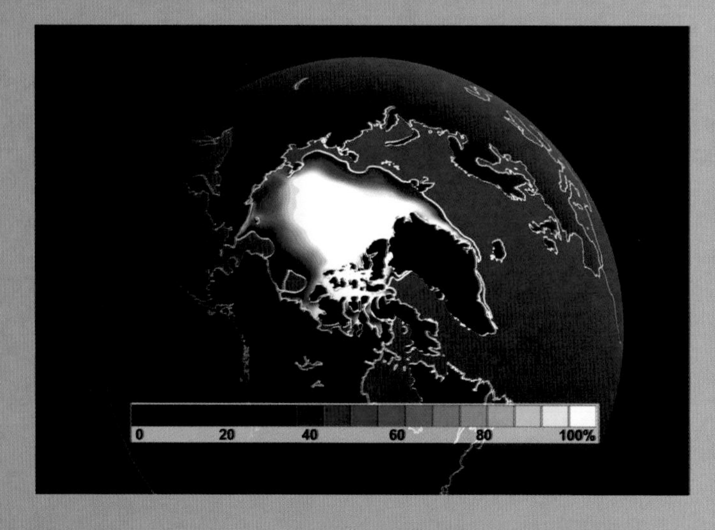

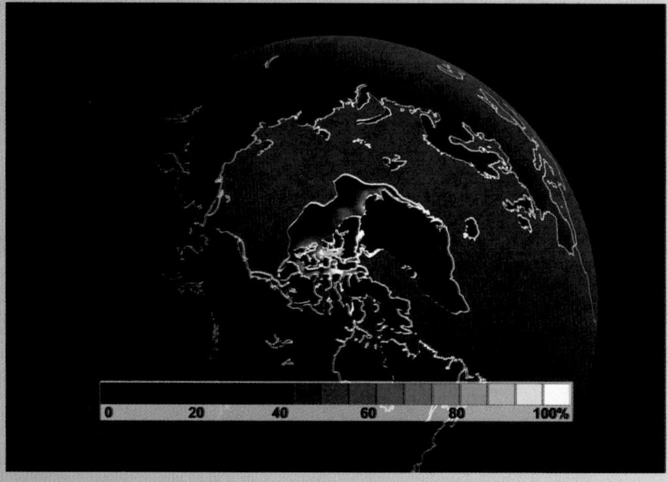

The arrival of spring in the Arctic is now taking place more than two weeks earlier than it did just a decade ago

Green Greenland

Mile long ice cores reveal that Greenland was covered with lush forests as little as 450,000 years ago

above: Researchers examine an ice sample from Greenland [image courtesy of Eske Willerslev]
opposite: Image showing the opening up of the Davis Strait between western Greenland and Baffin Island, Canada, during the summer months. Snow cover is making its brief, summer retreat from the west coast of Greenland (right), exposing the rocky landscape [image courtesy of Jacques Descloitres, MODIS Rapid Response Team, NASA / GSFC]

The oldest DNA yet recovered and analysed has shown that Greenland was covered with lush forests as little as 450,000 years ago. The samples were found at the bottom of cores that were drilled more than a mile (2 km) through the ice now covering Greenland. Wide varieties of trees were shown to have been growing there and include pine, yew, alder and spruce. Samples included genetic material from several species of invertebrates, such as butterflies, moths, flies, beetles and spiders. Professor Eske Willerslev of the University of Copenhagen in Denmark led the research and said the DNA is thought to date back to between 450,000 and 800,000 years ago.

Temperatures at the time the boreal forests covered Greenland are thought to have been about 10°C (50°F) during the summer and -17°C (1.4°F) in winter, with the reduced ice cover meaning sea levels were three to 6.5 feet (about 1 to 2 m) higher than today. Until the samples were taken, the most recent evidence of boreal forests on Greenland dated from 2.4 million years ago. The trees identified by the DNA testing grew during an interglacial period, which, when it came to an end, resulted in ice covering the region once more.

The ice cores further revealed that during the last interglacial period 116,000 to 130,000 years ago, when Greenland's average temperature was 5°C (9°F) warmer than today, the ice still covered the region. Professor Willerslev said that if the core data is accurate, it suggests the Greenland ice cap was more stable and less sensitive to climate change in the past than previously thought.

If man were to vanish overnight, his legacy would be a mere flicker in geological terms. With a few exceptions, all signs of his existence would disappear within about 50,000 years

The fall of man

Since mankind first settled down to cultivate fields and live in permanent communities instead of roaming the world as a hunter-gatherer, he has had an enormous impact on the landscape. Huge areas of the planet have been moulded to fit man's purposes, whether by building immense cities or turning forests and grasslands into fields. If he were to vanish overnight, however, his legacy would be a mere flicker in geological terms and it is estimated that, with a few exceptions, all signs would disappear within about 50,000 years. Radioactive waste would remain dangerous for longer, probably for two million years, and some manmade chemical traces would last for 200,000 years. But, essentially, there would be nothing left except for some fossil remains and a few brick and stone archaeological ruins largely buried beneath the surface.

Noise pollution would be the first to go, with roads and factories becoming silent immediately, quickly followed by light pollution—almost all of it gone within 48 hours as power plants, starved of fuel, ceased supplying electricity to automatic systems. After three months there would be significant reductions in air pollution from nitrogen and sulphur oxides, and in 10 years methane produced by man's activities would vanish from the atmosphere.

The disappearance of man would mean an immediate reprieve for most of the 16,118 plants and animals listed in the World Conservation Union's (IUCN) red list as under threat of extinction. Fish stocks would be expected to recover within half a century, by which time all but the most long-lasting pollutants would have disappeared from freshwater systems. Wildlife would start to encroach on manmade structures straightaway and within 20 years most minor roads and small settlements would vanish under vegetation. In bigger towns and cities it would take 50 to 100 years.

Buildings would begin to crumble quickly with no one to keep them in good repair. Wooden structures would decay within a century, and metal and glass buildings would collapse within about 200 years, by which time even the vast grain belt in the US would have returned to prairie. Few bridges would last more than 200 years, and the majority of the world's dams would collapse within about 250 years.

Climate change caused by greenhouse gas emissions would continue for another century, but after 1,000 years carbon dioxide levels would be back to preindustrial levels and all traces of manmade carbon dioxide would be gone after 20,000 years. Organic material buried in landfill sites would be almost all decayed within 1,000 years and in 50,000 most glass and plastic would be degraded.

opposite: The Chernobyl disaster resulted in deserted areas such as this, the ghost town of Prypyat, perhaps resembling what cities across the globe might look like if humans vanished overnight [photo: Oleksiy Shybanov / Ukrinform / UPPA / Photoshot]

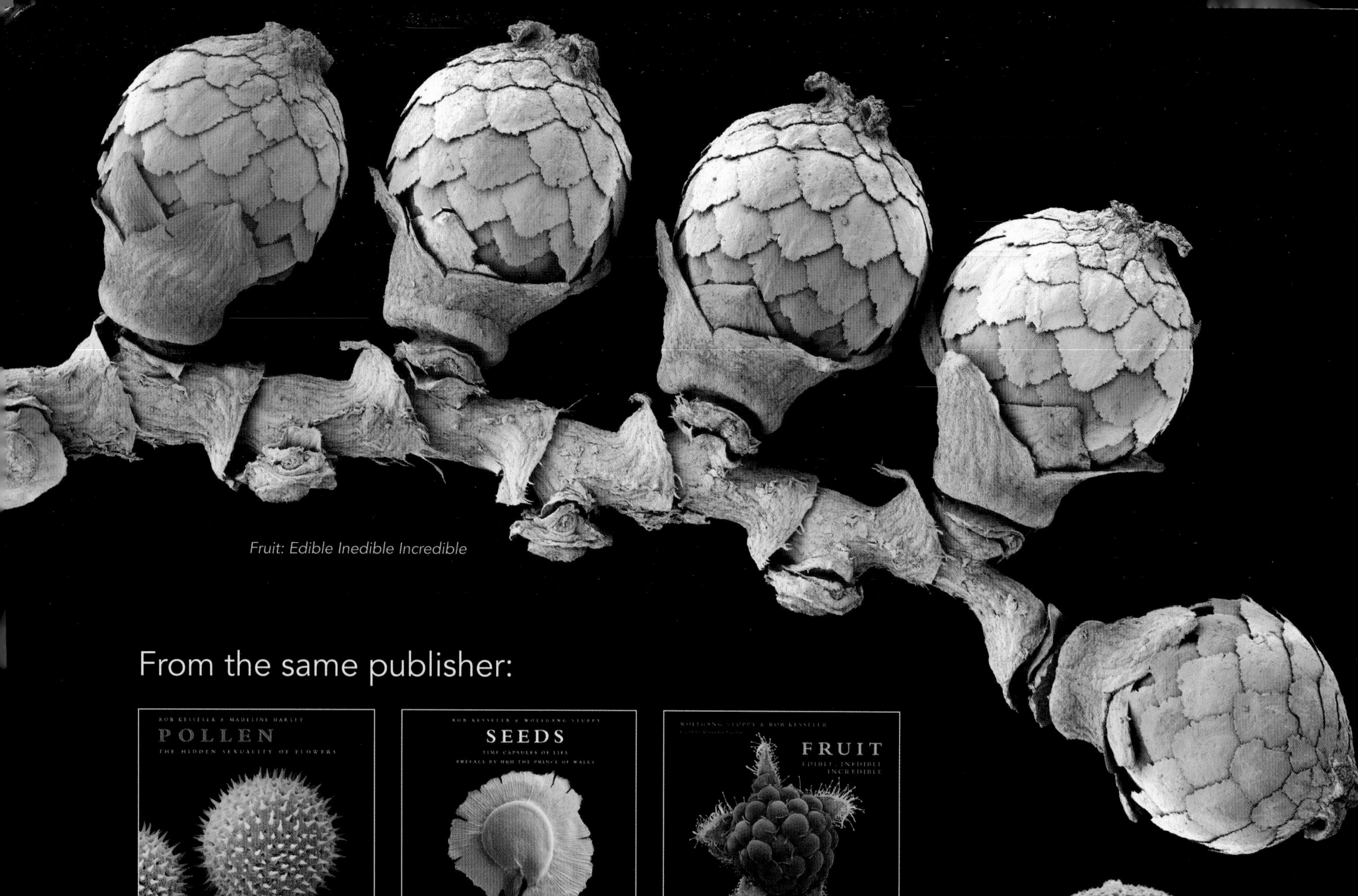

Fruit: Edible Inedible Incredible

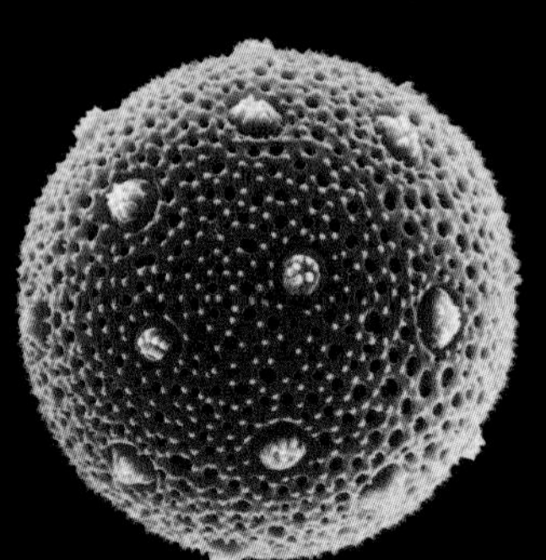

Pollen: The Hidden Sexuality of Flowers

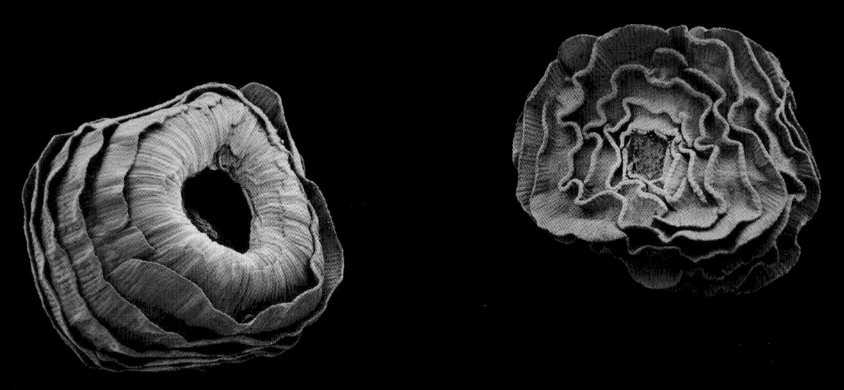

Seeds: Time Capsules of Life

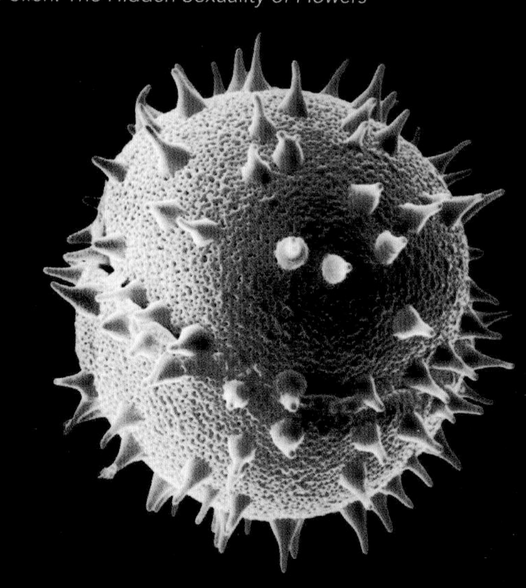

Kew

In collaboration with PLANTS PEOPLE POSSIBILITIES